Courage in Camouflage

A collection of true stories told by some of the country's finest military combat veterans

John H. Clark III

Cover and interior design by Dust Jacket Designs
Proofreading by Cheryl Williams

All stories included in this book were first published as a weekly series in the Killeen Daily Herald newspaper, beginning in July 2020.

Dedication

This book is dedicated to all the men and women who wear or have worn a uniform and served as a member of the United States Armed Forces.

As a whole, these brave souls make up the finest and most powerful military force the world has ever known.

As individuals, they represent a diverse cross-section of America, and a collective group marked by bravery, pride, honor, and dedication. Each person who swears to protect their country and give up their life if necessary deserves the utmost respect, heartfelt thanks, and never-ending gratitude.

Hoo-ah!

Contents

Introduction

When I first moved to central Texas from my hometown Houston 30 years ago to work as a reporter for one of the local newspapers, one of the area "beats" that eventually became my day-to-day responsibility was covering stories from Fort Hood, one of the largest military installations in the world and home of the historic 1st Cavalry Division.

Growing up in what is now the fourth-largest city in the country, I was never around much of anything military. My uncle Howard served as a crew chief on a B-17 bomber during World War II, but he never talked about it, and so my only personal experience with the Army and soldiers was watching the evening TV news reports on the Vietnam war, when I was a kid growing up in the late '60s and early '70s.

For a long time, I thought what I was watching on the nightly news was my ultimate destiny. That awful war and all those body bags were on television every day, and I just figured that when I turned 18 years old, I would be drafted into the service and sent to Vietnam, where I would die.

That is what I grew up thinking.

I don't remember being unusually upset or traumatized by that notion, and never talked to my parents or my friends or anybody else about it, but I remember thinking that is what eventually was going to happen.

Of course, the U.S. stopped its military draft in 1973, the year I turned 16, and the war ended a couple years later.

But I digress …

Even today, most non-military people outside of central Texas are not familiar with Fort Hood, which is always at least a little surprising, since the 214,000-acre post that houses 45,000 U.S. Army troops is far and away the driving economic engine for the entire region halfway between Austin and Waco, a fairly significant portion of the state. Several counties and dozens of small towns depend almost entirely on revenue generated by the local military for their economic survival – Killeen, Harker Heights, Copperas Cove, Gatesville, Nolanville – and so everyone in the area is well-acquainted with the installation that was founded during World War II.

It was not until I started living in the area and reporting on military news that I got to know some soldiers personally and found out that they are actually real people, and not just some kind of robots in matching, funny-looking uniforms.

One thing I always say about our soldiers is that they are just regular folks who do really brave things under incredible circumstances.

I have written hundreds of stories about the military and Fort Hood during my journalism career, and I am still writing stories about them today. I admire and respect our soldiers and appreciate very much what they do. Forget for a moment the "industrial-military complex" and all the politics – when you get right down to it, our armed forces are mostly comprised of regular, everyday people who just happen to have a job that requires them to run *toward* machine gunfire, explosions and bomb blasts.

Going to work and doing their job means they could die, and for the most part, these incredible men and women are always ready, willing, and able to do so. They enjoy what they do and are proud and eager to do it.

Sure, there are knuckleheads in the military, just like anywhere else. But mostly these are honorable human beings who truly want to serve their country. Of course, the prospect of a lifetime pension, medical care, and other benefits are motivating factors, but most all of the soldiers I have met over the years are upstanding, squared-away patriots who believe in what they are doing.

Hence, the reason for this book.

My first three decades on this Earth, I never once talked to a soldier. Never had the need or opportunity. Sure, I probably met a soldier here and there, but – aside from my late uncle – never really knew anyone who wore the uniform. And unless a person lives in the vicinity of a large military installation, or has a family member in the service, chances are good that they are in the same boat.

So, I present to you a collection of stories about some of the finest men and women who have served in the United States Armed Forces.

All these profiles first appeared as a once-a-week series in the Killeen Daily Herald newspaper. Special thanks to KDH management and editorial staff for supporting this effort to honor some of the unsung heroes who make our military and our country the greatest in the world.

Each story remains exactly as it originally appeared in the newspaper.

Let me introduce you first to…

Scooter Barclay

Veterans of World War II, also known as "the greatest generation," are a vanishing breed with a reported 389,292 of the roughly 16 million Americans who served from 1939-45 still alive, and 370 of those dying every day.

Kenneth "Scooter" Barclay is one of those heroes of history still alive and well.

Now 96 years old, the Spokane, Wash., native enjoyed a successful real estate career in central Texas after the military and lives now in an assisted-living facility in Copperas Cove, Texas. He sometimes uses a wheelchair to get around, drives his own pickup – or used to, before the COVID-19 pandemic put restrictions on his independence – and plans to be around for at least a few more years.

"I want to make it to a hundred," said Barclay, a decorated former U.S. Army Special Forces (Green Beret) member who retired in 1973 as a lieutenant colonel after 30 years' service. "I would like to do anything I can to get people who make it to a hundred recognized."

The father of two, grandfather and great-grandfather volunteered to be drafted when he was 18, fought in the south Pacific, got out of the military for a while after the war, went back to school and earned a degree in political science and economics, then re-joined the Army.

"I enjoyed the Army, and just always thought I would go back," Barclay said.

He was stationed at Camp Hood (now Fort Hood) in 1949 and met a local girl on a blind date. That girl would later become his wife, but things did not get off to a flying start that first night.

"At that time in Texas, liquor laws were by city and by county," Barclay explained. "Killeen (adjacent city) and Fort Hood were completely dry. They didn't have a liquor store at Fort Hood, (and) 3.2 (percent alcohol content) beer was all you could get. If you went to Temple, you could drink beer and what-not.

"There was this place across from the (military) hospital in Temple called the Idle Hour. We went to this place, and, apparently, I drank too much. Well, I'd been brought up to be a gentleman, so I took my (future) wife home, walked her to the door; she got inside, shut the screen door, and said, 'Good night. I don't ever want to see you again.'

"I thought, 'Man, I'm a better guy that that.' So I called her and prevailed on her for another chance. I said I was sorry; I got out of bounds; I'm a better guy than that; let's try it again – you know.

"So we went out again, and we saw each other for about a year. I'd always wanted to go airborne in World War II, and so when I finally got orders to go airborne, I asked her to marry me before I left. I went off to six or eight weeks of jump school, then on to Fort Campbell. Every Friday, I'd go to bingo, and I'd call her after bingo, and ask her to marry me.

"She finally surprised me and said, yes, so we got married in 1951."

Barclay, who lost his wife, Jacqueline, to cancer after 60 years of marriage, says he is sometimes a little reluctant to talk about his military career because he does not want to be seen as tooting his own horn. But he thinks it is important that the sacrifices of veterans like himself and his buddies are never forgotten.

"People who do brave things aren't brave when they do 'em," he said. "They're doing something because they're in a situation where suddenly, it's kill or be killed."

"I've known several Congressional Medal of Honor members (the military's highest award for bravery in combat), and they all told me the same thing. (Lloyd) Scooter Burke told me – I knew Scooter before he got (the honor) and after he got it – he was in Korea, they were facing the enemy and they weren't doing very well. He said, 'I can't stand it anymore,' and he grabbed a machine gun, jumped up and started firing. The guys who were with him saw him do that, and they joined him.

"He didn't start out to be a hero. He just said, '(Screw) it, I can't stand this anymore,' and he did something."

One of the most intense battles he participated in during WWII came on the island of Saipan, considered by the Japanese to be part of the last line of defenses for the country's homeland. That fight, regarded as one of the major campaigns of the war, lasted from June 15 to July 9, 1944.

When it was finally over, 3,426 Americans were dead, and 10,364 wounded. Of an estimated 30,000 Japanese defending the island, a total of 921 were reportedly taken prisoner.

"When we went to Saipan, my infantry company had 188 men. When we got out of the line, we only had 48 people who were still there. Now, about half were killed or badly wounded, (and) we had a lot of heat casualties, where people had to be evacuated and taken to hospitals.

"I can remember some Japanese coming out to try and surrender," Barclay says. "One of them was literally naked, carrying a bamboo pole with a bayonet on it. The others might have had a weapon, but I don't think they did. A guy said, 'You're not taking my place on a boat back to the States,' and he went bang, bang, bang, bang, and killed all four of 'em.

"Now, that's an atrocity, but it wasn't in World War II."

Along with the Bronze Star, Barclay says he apparently also was recommended for a Silver Star – the military's third-highest award for bravery in combat – but never received a medal.

"On Okinawa, I got involved in a situation where I killed a bunch of (enemy troops), got up on top of a tank and led the tanks firing, and I got recommended for the Silver Star. My understanding is that it was approved, and one of the reasons I believe that is, one of the people who was in my company told me he saw the orders, but at the very end of the war, they had a big typhoon and a cruiser went down, and also a ship that carried records.

"My belief is that my records were on there. So, you know … I know what I did, and a Silver Star and a dollar will buy me a cup of coffee."

Some of the things he saw and did during the war haunted him pretty severely when he got home, Barclay says, and on occasion, continue to bring him pause even today.

"One time, I was visiting at a school, and this kid looked at me – he was about 10 or 11 years old – and he said, 'Did you ever kill anybody?'

"I think it was the first time I ever talked to a group of kids, and I thought, 'What the hell am I supposed to say?' I decided to tell the truth, so I said, 'Yeah, I did, but you have to understand that the purpose of war is to win, and to win, you kill or be killed. If you don't kill the enemy, they're going to kill you.'

"After the war, when I came home … I used to have bad, bad nightmares. I would get up in the middle of the night, screaming and fighting and carrying on. My mother would run in and shake me and ask what was going on. That went on for maybe a year, I guess.

"Today, once in a while, I'll think about it, because I'll remember somebody (long pause) …

"I've been blessed. I had a very interesting Army career; did all kinds of interesting things. The Lord has taken good care of me, that's for sure."

Homer Garza

Retired U.S. Army Command Sgt. Maj. Homer Garza grew up picking cotton and other cash crops with his eight brothers and sisters in small-town South Texas.

Working on the family farm during World War II, he decided as he got a little older that joining the military sounded like a good way to broaden his horizons, and so he signed up for the National Guard in 1948. After two weeks' training at what was then called Camp Hood, 16-year-old Garza volunteered for active-duty and went on to a 25-year career that included combat duty in two wars and becoming one of the most highly decorated soldiers in central Texas with awards including two Purple Hearts, two Silver Stars, three Bronze Stars with V device for valor, a Legion of Merit, Army Commendation Medal with V device.

"The worst (part) of my career was Korea in 1950-51," Garza, 88, said, recently, from his home in Harker Heights, Texas. "It was really rough. When we left Yokohama, Japan, in July of 1950, we had about 250 people in our company. I rotated in May of 1951 and there was only 11 of us left.

"I was wounded twice. I took some shrapnel in my leg in August 1950, and some hand grenade shrapnel in January 1951 by the Chinese. The first time, I went to the hospital in Pusan and was there for three or four weeks, then returned to my unit. Still had bandages on my leg. When I got wounded a second time, I never left my unit.

"When you're in that situation, you want to go back to your unit. You're thinking about your fellow soldiers. You don't think so much about your country – you think about staying alive and protecting your buddies."

In 1968, Garza was sent to Vietnam, where he served one year as a first sergeant in an artillery unit during the famed Tet Offensive, one of the largest campaigns of the war with a series of coordinated attacks by the North Vietnamese on more than 100 cities and outposts in South Vietnam. News coverage of the massive Tet campaign and its heavy casualty toll marked a major turning point in the war, as support for the effort began to erode back home.

Garza's unit provided artillery support for 9th Infantry Division soldiers on patrol. Much of the time, those artillery crews traveled on boats and barges along various rivers. In some places, thick jungle was only feet away from the barges, which were armed with two 105 mm howitzers.

One day, Garza was hit by shrapnel from enemy artillery, slicing open his back. A doctor who stitched him up asked Garza if he wanted him to put in the paperwork for a Purple Heart.

"No, I already have one," Garza said.

After Vietnam, he came back to Fort Hood for a while, then served in Germany in 1971 as command sergeant major of the 3rd Infantry Division and retired two years later from Fort Hood.

When he left the military, Garza went to work for the city of Harker Heights as a building inspector and later fire marshal. He started his own business as a home inspector in 1983, sold the business in 2000 and retired for good.

During his military career, Garza spent much of his at-home time stationed at Fort Hood. His first duty here was in 1949, and he remembers how much things have changed since then.

"At that time, there was no Harker Heights," he said. "There was nothing in Killeen. When I got here, most of the streets were caliche and dirt. There were very few paved roads. All you'd see was dust from automobiles going down the road.

"There wasn't much to do. We would hitchhike out of the main gate and go to Belton and Temple. They used to have a big dance hall in Temple, and we went to the dances there. I couldn't drink beer because I was not 21 years old yet. We had older buddies buy a case and we'd sit outside the barracks and drink beer."

These days, the great-grandfather spends a lot of time at the house with his wife of 68 years, Hilda. He keeps busy with his self-assigned duties as a "remote operator."

"I operate the remote for the TV," Garza said.

He is troubled by what he sees going on in the world, but tries not to let things like the pandemic, protests, and riots get him down.

"I was in Germany in 1952 when the polio epidemic hit the United States, so I didn't see too much of that at all. My wife got there in 1953 and we came back in 1956 – went to Fort Riley, Kansas – with a brand-new little girl.

"I just block all those things out of my mind. I don't agree with what they're doing. If they went what I went through, they wouldn't be kneeling. They'd be standing at attention."

Rafael Riera

Former U.S. Army Spec. Rafael Riera is still haunted by memories of the April 4, 2004, ambush in Iraq, a bloody battle that left eight American soldiers dead and more than 60 wounded – the largest one-day casualty count since Vietnam for the famed 1st Cavalry Division.

"What happened that day basically lived with me for a very, very long time after the fact," said Riera, a 40-year-old native of Puerto Rico who served at Fort Hood, lived in Killeen for a long time, and moved last summer to Kempner.

Known as "Black Sunday," the fight for Sadr City (a large Baghdad slum with a population of two million people) became the subject of a non-fiction book, "The Long Road Home," by ABC News reporter Martha Raddatz, and a 2017 television min-series of the same name.

The mini-series was filmed at Fort Hood, using one of a number of MOUT (military operations on urban terrain) training facilities designed to replicate overseas towns and villages, including Sadr City. Soldiers who were there served as consultants to help ensure accuracy. Riera was also called upon to be an "extra," or background performer.

"Visiting the MOUT site was a rude awakening," he said. "We were part of the QRF (quick reaction force) team that day, looking for a downed platoon that was on a routine mission to provide security for sewage trucks.

"All hell broke loose and they had taken shelter inside a building in the middle of the fight. I never realized how close we were to finding them, but when I went down to the (Fort Hood) training area where they had built a mock-up of the city, it broke me down. I had to leave because I couldn't stop crying.

"It was just like being back there – minus the smell. You can't really imitate that Iraq smell. A combination of defecation, staleness, everything decomposing … it's a very unique smell.

"When I watched the show – reluctantly – it opened my eyes to a lot of things that I never even realized happened while I was over there."

Born in San Juan, Puerto Rico, Riera came to the U.S. at age six or seven when his mother went to work at the Pentagon. They returned home following a death in the family, and after graduating from high school and attending a year of college, Riera moved to New Jersey to live with his father.

He was living and working on the East Coast when the historic 9/11 terrorist attacks struck New York City, a devastating event that led him to join the Army.

"I was supposed to take my stepmother to the airport that morning," Riera said. "Turned on the TV and saw everything happening. Got into my truck – obviously the airport was not going to happen – and drove over to the shore and watched everything from the hood of my truck, right off the Jersey side of the Hudson (River).

"It was nothing but big grayish, white clouds … I can't even describe exactly how it felt to watch it. But it gave me the feeling that I needed to do something about it.

"Shortly after that, I joined up."

Riera was 22 years old when he reported to boot camp at Fort Benning, Ga. It was tough at first, but the young man worked hard and even earned a second stripe while he was there.

"It showed me how out of shape I really was, but I picked it up quickly. I did what I was told and actually got promoted in basic training (from E-1 to E-2)."

After completing weapons school, Riera was assigned to Fort Hood in the summer of 2002 as an infantryman with the 1st Cavalry Division's 2nd Battalion, 5th Cavalry Regiment. Two years later, he was on his way to Iraq and Camp War Eagle, on the outskirts of Sadr City, for a 14-month tour.

"I remember just being anxious to get there," he said, of the long overseas flight. "Nervous, of course. Everything you had trained up to do, now it's real life. Nothing can compare to the real thing.

"If anything, you become a true American; a true patriot. You want to defend and help anybody and everybody that needs help. You want to get there and put all your training to use, and it's only after you're actually there that you realize it kind of sucks, and you want to go back home.

"You never know how you're going to react to something like that (combat) until you're actually in it. There's a lot of people who did a lot of talking before we actually deployed, saying that they were ready and couldn't wait. As soon as bullets started flying, they just couldn't handle it. There were also people you thought were going to break down and not be able to handle it that turned out to be mini heroes of the day.

"I found out pretty quick that whenever I thought I was about to get shot at, I have a nervous twitch in my right leg. Other than that, I pretty much had a job to do."

After his deployment, it was back to Fort Hood for a short stay, then out to Hawaii for a few years. While he was stationed there, a serious shoulder injury he suffered in Iraq became too much and he was medically discharged in 2008.

"There was an incident where I was leaning up against the Bradley (Fighting Vehicle) ramp and an RPG hit exactly where I was. The concussion of it tore a ligament and pinched a tendon in my right shoulder.

"I didn't realize it because I was popping Motrin like Tic Tacs, and so the pain wasn't affecting me as much. When you're in the infantry, you get used to a little bit of pain and a little bit of constant ache here and there. When you go back to your regular training and your body relaxes is when it (injury) actually makes itself evident, and that was the case for me.

"I'd much rather be working, if I could give the pain back … I hate taking pills and so popping pain pills just so I can keep going day-to-day is not exactly fun."

Now fully retired due to his disability, Riera keeps busy with his duties as sergeant-at-arms for the Combat Veterans Motorcycle Association local chapter. Members of the group usually earn colorful nicknames somewhere along the way, and Riera is no exception.

"We all earn our names one way or another, or we choose our own – unless we do something stupid or unique. My name is 'Meat,' and I'm not going to go into details about that. It is not fit for newspapers.

"Our motto is Vets for Vets," said Reira, who rides a 2008 Harley Street Glide. "We are a group of combat veterans who help other veterans in need, and we just so happen to ride motorcycles, too.

"Originally when I joined the CVMA, my name was Fixit, because if it was broke, usually I could fix it. I don't know what it is, (but) I like to help other people more than I like to take care of my own thing. So I spend a lot of time helping other people – more than I do working on my own projects."

Looking back on his military career, Reira says he would do it all over again, although he might change a few things.

"I'll never regret the fact that I joined the military. I'd never have met my wife (Crystal) if I hadn't joined the military," the father of three said. "I might have chosen a different job – something that would have helped me in the long run get a better job after the military. But I don't regret joining at all.

"Don't sweat the small stuff. If you start worrying about all the small things a little bit too much, you'll miss all the good stuff."

Mike O'Hara

Mike O'Hara survived firefights, mortar explosions, and rocket attacks during three tours with the infantry in Vietnam, but the emotional scars and pain he brought back home nearly destroyed him.

"I drank an awful lot," O'Hara says. "Up until recently – maybe 10 years ago, when VA finally got a handle on what medications worked for me – I was drunk every night. Got into a lot of trouble over that stuff, but it's the only way I could deal with it.

"After the first tour, I leaned (that) you don't make friends – because then it hurts when they get killed. If you don't make friends, and you just have acquaintances, then it's just something that happens."

O'Hara went to Vietnam the first time not long after joining the U.S. Army in 1965. He was what some might call a troubled teen, raised by a single mother in Massachusetts, and when his rebellious ways landed him in serious trouble with the law, it was off to boot camp.

"I have to admit, I was a pain-in-the-ass. I was not a good child in my teens – staying out late, drinking, basically raising hell. What finally happened was I got caught joyriding (in a stolen car) with a buddy. My mother said, 'That's it.'

"She went down to the judge and said, 'My son's in trouble; what can we do?' So I ended up going into the Army, two days after I turned 17."

Even after he put on the uniform, young O'Hara had trouble staying out of trouble. It didn't take him long, though, to figure things out.

"I learned fairly quickly that if you keep your (stuff) together, you don't get yelled at as much. I did a lot of pushups until I started to learn to keep my mouth shut."

His first duty station after boot camp and advanced training was in Hawaii, with the 25th Infantry Division. Six months after he got there, it was off to war, and a life-changing three years of combat.

"The first tour was continuous patrolling. And I'd say if I learned something from that time, it was the utter tiredness. You're tired all the time. We went on Operation Paul Revere – which was actually four operations – and I heard it was the longest continuous combat operation to date: 249 days. Our battalion was out on all four of them.

"When you get shot at, it's actually a relief, in a way. All the other time, you're waiting, and all that nervousness is there. Then, finally, when it happens, you can do something back," O'Hara explained. "It was very surprising, because it did not make the noise I thought it would. The sound you hear in the movies? No, it's not like that – it just 'snaps.' A real sharp snap. It's hard to describe. It's that breaking of the sound barrier when it goes by you. A machine gun is a whole bunch of rapid snaps.

"When you get shot in the head, it's not like in the movies, either. You don't stand there. You just fall, like a bowl of jelly. One time, I was running right behind a guy that happened to.

"The one that shocked me the most – he wasn't a close friend, but he was a friend I knew – was a kid, and he got a bullet crease right through here," O'Hara said, pointing to his bicep area. "It took about an inch of skin off. And he was sitting there (saying), 'What's my mother going to think? What's my girlfriend going to think?' And then he died of shock. He just went into shock and died."

After a second tour with a Mobile Advisory Team, and a third tour as an advisor, O'Hara came back home and spent another 17 years in the Army, keeping his nose mostly clean and retiring as an E-8 (master sergeant/first sergeant) after 20 years' service. He lived in Harker Heights then, and went to work on an offshore oil rig for two years, then managed a 7-Eleven store, attended classes and taught part-time at Central Texas College, before the VA declared him 70-percent disabled and unemployable.

Now 71, O'Hara has a house in Lampasas, about 30 miles west of Fort Hood, and spends his time making custom knives and hanging out at VFW post no. 8539, where he serves as house committee chairman. He is happy with his life, and says he has no regrets; never looks back and wonders what might have been.

"I would say, my life is really no different than anybody's. I've had my ups, my downs, maybe some specialized problems.

"I've never been married. I think I came close probably twice, but I knew I wasn't ready for it. I didn't think they deserved the crap that I had going on with me. They were good girls – it might have worked out; might have not. By the time I figured out it might be a good thing to be married, I was too damn old," he said, laughing.

"I came out of 'Nam not really caring about people. I'm indifferent to people. If someone dies, I'm like, 'OK, well, see ya.' It's nothing I mean to do; it's just the way I am. Now, I've never done anything violent. If I was going to say I was dangerous in any way, it was to myself.

"I very rarely drink now. If I drink to any extent, it's at my house. If I come up here (to the VFW) and drink, I'll have maybe two or three beers, and I usually time them, one an hour. I'm paranoid about drinking and driving. I will not do that. If I ever get caught again, it'll be straight to jail.

"A lot of my troubles came after I retired from the Army. When you're in, you're too busy to think about those things … but I don't blame what happened to me so much, the problem was how I handled it. A lot of soldiers are too damned proud to go ask for help, and I was like that."

Once a soldier, always a soldier, O'Hara says. He is proud of his service and would do it all over again. Vietnam was a controversial war, and the first one considered "lost" by the United States, but he still believes it was the right thing to do.

"I have got to say that it was just, because I lost too many friends over there to say it wasn't. I think we were right. The military did not lose the war over there; the politicians lost it. I'll tell you right now – the biggest thing we have to fight in the Army is not the enemy, it's the politicians back here that makes these rules of engagement," said O'Hara, whose military awards include the Silver Star, Bronze Star with 'v' device, Purple Heart, and Army Commendation Medal with 'v'.

"I am a strong believer … you've got light colonels and colonels, with 20 years in the military, training nothing but their specialty. Then, you've got a first-term representative come up and tells them what to do. They get people killed.

"I am a firm believer that when you let loose the dogs of war, you take that whole collar off. Just give 'em their objective and say, 'Take care of it.'

"I'm a soldier, and I will always be a soldier. I believe strongly that when I took that oath of office, it never runs out. I will always protect the Constitution. I will fight, even at my age, to protect it against all enemies, foreign and domestic."

Stacey Ochoa

Stacey Ochoa and her husband, Danny, named their new baby boy, Maverick, but the couple insists it had nothing to do with the fact that mom serves as a technical sergeant with the U.S. Air Force Reserve.

"Top Gun (the famed fighter pilot movie starring Tom Cruise) is a reference that a lot of people make," Ochoa said, laughing. "But really, what we wanted was a strong 'M' name – to go along with my 10-year-old daughter, Madison – that not many people have. So we decided on Maverick."

Ochoa, 31, is the daughter of a career soldier, and spent her formative years in Killeen, Texas, graduating from Ellison High School in 2006. She worked for a while after high school as a waitress, and then on the boat dock and administrative offices at Belton Lake Outdoor Recreation Area on Fort Hood. After her first marriage ended, she decided to make a major change and follow her dad's and brother's footsteps into the service.

"Growing up in the military, you're surrounded by it all the time, so it was something that was always in the back of my mind, but it wasn't something I thought that I'd be able to do," she explained. "Then, shortly after I separated from my husband, I decided I wanted to do something for myself. I was kind of stuck in a job that I didn't really like, and wanted to broaden my horizons, if you will.

"My dad was in the infantry, and he told me not to go Army. My brother had joined the Air Force in 2010, and he told me it was probably the best decision that he ever made, so I figured, why not?

"I signed my contract in 2011 and left for basic in 2012."

As a Reserve, Ochoa normally reports for duty and training with the 73rd Aerial Port Squadron in Fort Worth, one weekend a month, and two weeks a year. Like a lot of other part-time troops in recent years, however, she has also spent her share of time on temporary active duty during deployments to the Middle East. Last October, she got back from her second overseas tour, where she was stationed in Iraq.

"My job specifically is aerial port. We load and download aircraft; inspect for hazardous materials, make sure it's fly-worthy, chain it down inside the aircraft; make sure everything is weight-distributed correctly. This last deployment, I was in charge of inspecting the cargo.

"There were 12 of us that went, and I was the only girl, so I was 'Mom' for everybody, which was pretty fantastic," Ochoa said. "Three of our guys, this was their very first deployment, so it was really cool to be able to help them, since the rest of us kind of knew the ropes.

"It was a very small port that we worked at – Erbil. Everybody had to come through us to get their stuff shipped back home, or to other locations. We worked in six-man teams – six during the day, and six at night. It's the busiest (aerial) port in Iraq, so we were just rockin' and rollin'.

"I was the mama, but I also had a bunch of big brothers, so I always had someone looking out for me, which was really nice."

That was a busy mission, but the trip to Afghanistan back in 2014 was quite a bit more dangerous.

"Afghanistan was definitely a lot more active than Iraq was. We had four air attacks while we were there, and two ground attacks," Ochoa said.

"The first time was a ground attack. You don't know exactly what is coming – are they coming on foot, or in vehicles – so it's a matter of listening to the radios and trying to figure out what's going on. That time, it was four kids who broke the (perimeter) barrier and came in. Essentially, it was just kind of a recon mission to see how far they could get in before somebody saw them, or something happened.

"Another time, we had a rocket hit the flight line – which is where we worked – a little bit down from our building, and that shook everything.

"You just kind of hunker down and wait for it to pass. We're the Air Force, so we just kind of wait for the Army to do their job," she said, smiling. "You put on all your equipment, go down to a bunker, and just kind of hang out and wait for the all-clear.

"It was scary, but at the time, it's also kind of exciting. Once you have time to sit back and think about it, it's pretty terrifying."

Ochoa's husband, Danny, is a staff sergeant with Fort Hood's 1st Cavalry Division infantry. He has been deployed four times to Iraq and Afghanistan, beginning with the 2003 invasion of Iraq, so he knows all about bombs and bullets flying. Part of his job was to help clear the way for convoys moving through the desert, considered one of the most dangerous missions.

"Bad guys are going to do bad guy things, and you just have to make sure you don't give them that open window to do what they do," the 37-year-old California native explained. "They're going to try and find a way – sometimes we catch them, and sometimes we don't.

"I got back home after that first year, and three months later, we went back to Afghanistan. Came home for about a year, then went back to Iraq, and then back to Afghanistan. I actually enjoyed it."

Stacey explains:

"After you do the first one (deployment), you get this 'deployment itch.' You don't know what it is until it hits you," she said. "It's a simpler way of life. You have two (concerns): doing your job and staying alive – that's it. That's your main focus, day in and day out. Then you come home and there's the redundancy of work, family responsibilities, money …

"You get back home and get back into your normal routine, and then after a while, you go, 'I could really go for another deployment right now.'"

With two years left on her current commitment, Ochoa – who has a Bachelor of Arts degree in English, and also works as a substitute teacher in the Killeen school district – not only plans to stay in the Air Force, she is thinking about going full-time, active-duty. That would be around the same time Danny reaches the 20-year mark and retires.

The couple, whose second wedding anniversary comes next March, met on a Saturday afternoon in Salado. Talking about that day brings a smile to both their faces.

"I didn't have my daughter with me, and I wanted a nice, relaxing day, so I went to a brewery over there and had a beer and was reading a book," Stacey said. "He had just finished golfing, and he walked over to try and figure out why I was reading a book in a bar."

Danny says it took a while to gather his courage, but he finally decided to make his move.

"I mustered up the courage to go over there and talk to her," he said. "I think I asked her what she was reading. I sat down and we talked for a while, then we exchanged numbers, and I texted her a couple days later."

Another thing that brings a smile to Stacey's face is a Veterans Day tradition she shares each year with her dad.

"It's funny, the first thing I'll get in the morning is a text from my dad asking if I've gotten my free breakfast yet," she says. "He's up at six o'clock going to Denny's or IHOP, or whatever. I'm, like, 'Dad, I haven't even gotten out of bed yet.'

"I look forward to those text messages every year."

Jim Reed

Salado resident Jim Reed, executive director of the Central Texas Council of Governments (CTCOG), flew Apache attack helicopters in the U.S. Army, and says piloting those powerful tank-killing machines sometimes got him into trouble on the ground.

"My wife will tell you that I used to drive like I flew," Reed said.

"She used to always say, 'Jim, this is not an Apache.' It changes your perception. When you're used to the speed and sensation and adrenaline of flying on the deck at night at a hundred miles an hour or more, then you get in your car and drive home, it seems real slow at 70 miles an hour."

Reed, who grew up in Fort Walton Beach, Fla., with a sister and two brothers, played football in high school, ran track, was a member of the marching band and sang in the choir. He went to the University of South Alabama on an Army ROTC scholarship, where he was a distinguished military graduate and thereby earned a regular Army commission.

"I graduated on May 22, 1985, and was commissioned the following day," he said. "With everything we owned in a U-Haul, we drove from Mobile, Alabama, to Fort Rucker, and I reported for duty."

Based on his finishing at the top of his flight school class, Reed was chosen to participate in a new program as the Army transitioned from Cobra helicopters to the new AH-64 Apache attack helicopter. He was one of only six recruits (lieutenants) designated to field the Apache at Fort Hood.

"In 1986, the Army was in the process of bringing the Apache helicopter on board," Reed said. "The Apache was very high-tech – still is – and at that time, computers weren't what they are today, and so the Army decided to come up with an experimental program. What they wanted to do was pick six of the Army's newest aviators to see if it was possible for someone with very limited flight hours but a proven aptitude – you had to be number one in your flight class – to be able to handle going straight from flight school into the Apache.

"Up to that point, you had to have a lot of experience to be one of the first Apache pilots. You had to have over fifteen-hundred total (flying) hours, and some combat experience, because we had Vietnam veteran pilots at the time.

"They called us 'the first turnaround class,' because it was the first group that turned around directly from flight school and went to the Apache. Now, it's commonplace, but we were the pilot group."

Reed describes some of the intricacies of his Apache training at Fort Rucker, the Army's primary flight training installation:

"You have to learn all about the aircraft, which is highly technical, and how to employ it, which is highly tactical. But you also had to learn how to fly at night with no lights, using a forward-looking infrared radar system.

"To teach you to trust that, you fly in the daylight hours, but what they do is they seal off the cockpit with a black curtain, and they call it 'the bag.'

"The front seat pilot is the instructor pilot, and he's sitting in the regular cockpit, so he can see it's daylight. You're sitting in the back, basically 'in the bag,' and you have to learn to fly the aircraft utilizing just the instrumentation the computer and the camera gives you – and that's all projected onto your right eyeball.

"You have to fly using the infrared picture, and the computer puts certain flight instrumentation on top of that, so while you're looking at that camera with your right eye, you're (also) seeing your air speed, altitude, targeting, power settings of the aircraft, and all of the instrumentation.

"It's a real challenge, because you have to learn to trust that picture you're seeing, and it also causes some problems for you because the picture you're seeing is actually 15 feet in front of you. That's where the camera is. So when you look to the left – when you're not in the bag but flying normally – your left eye picks up the outside cues, and your right eye picks up what is 15 feet in front of you.

"When you look to the right, your left eye picks up the visual cues, but your vision crosses, because your right eye picks up what is 15 feet in front of the aircraft. Your left eye sees what is right there, but your right eye sees what is fifteen feet out to the left.

"You have to learn to disregard the information that your brain is used to getting from your eyesight, and a hundred percent trust what the helicopter system is telling you – and you're doing this at 200 miles an hour, ten feet above the ground, at night.

"The Army was concerned as to whether someone with limited piloting time could handle that overload, and it turned out that you could."

After his training, Reed reported to Fort Hood, which at that time was the only Army post with Apaches. He spent the rest of his five-year career with the 6th Air Cavalry Combat Brigade, preparing for war but never flying into combat.

He left the military as a captain after five years' service and returned to Alabama to attend graduate school at Auburn University.

"At that time, the Gulf War was going on, and although I was off active duty, they did not know whether they were going to need me back. I was in the Reserves under hip-pocket orders to be ready to return, should they need me," Reed said.

"My father-in-law was a professor at Auburn University, so I went to grad school there, so that if I had to go back into the military, my family would be near family. We had our second child at Auburn, and then I went to work for an engineering firm for a while, and ultimately returned to central Texas in 1993, to go to work for the Council of Governments."

Although he was a combat pilot who never saw combat, Reed says he is proud of his military career, and particularly his historic role in helping field the Apache, one of the Army's fiercest weapons that carries two crew members, a pilot and a gunner, with a combination of Hellfire anti-tank missiles and 70 mm rockets, and a 30 mm single-barrel, chain-driven autocannon.

"I always wanted to fly, and it was a tremendous honor, but also a tremendous challenge," said Reed, a 57-year-old father of two who has been married to Lynette for 36 years. "To me, it was exhilarating. You're in the most advanced helicopter in the world; you're one of six lieutenants who is getting the opportunity to fly it. The last thing I wanted to do was be bad at it and fail.

"You had a lot to live up to. My first unit, the average amount of flight time was over three-thousand hours, and I came out of flight school with two hundred. So here I am, the lieutenant who is supposed to be platoon leader for these highly accomplished, best-of-the-best aviators – most of them combat veterans – and here I am coming in to lead them into battle.

"It is the highlight of my career and probably the thing I'm most proud of. In 1987, when an Apache flew over, people stopped. Even to this day, if you tell someone you're an Apache pilot, that's all they want to talk about. They want to know all about it, even though it was thirty years ago.

"I enjoyed the hours that I spent in the air. Did some aerial gunnery work up near Gatesville – shot a lot of rockets and a lot of 30 mm cannon rounds. That's always fun.

"I was extremely fortunate. I served until the beginning of the Gulf War. A lot of guys that I trained, and trained with, went and fought that war.

"I went on terminal leave the day before Kuwait was invaded. I was the deployment officer for 6th Cav Brigade when I left the service, and so as a young captain I went on terminal leave, and the phone rang. They said, 'You need to come down to the brigade headquarters. We need to understand some of these go-to-war plans that you were author of, because we may be going.'

"I was honored to be prepared to go, and I honor those who have gone and served during time of war. But I do consider myself blessed that I did not have to take on that challenge."

Angel Valencia

As he looked forward to last month's 82nd Airborne Division reunion at Fort Bragg, N.C., retired First Sgt. Angel Valencia said his buddies were telling him to get ready for some all-night celebrating and no-telling-what-all former paratrooper carousing.

"They were saying, 'Yeah, man, I'm so glad you're coming. We're gonna get some kegs; we're gonna party all night long, like we used to ...

"I said, 'Dude, come on, we're going to be putting on our CPAP (night-time breathing machine) and laying down at 10 o'clock – I know I am. Those days are long gone. I'm not 22 anymore."

Valencia, a member of the elite Sgt. Audie Murphy Club who now serves as a local church pastor, may not be staying up all night anymore, jumping out of airplanes, or humping through pitch-black jungles in nighttime training missions, but the 65-year-old father of two and grandfather of three from Copperas Cove remembers those old days fondly, and never gets tired of talking about them.

"Oh, yeah, the 82nd Airborne has an awesome reputation, and I'll always be proud of that. I always wanted to be the best."

Valencia grew up in Santa Barbara, Calif., the oldest of nine children raised by a single mother. He dropped out of high school when he was 16 to help support the family, and quickly climbed the ladder in the local restaurant business, making it up to manager after one year. Unfortunately, he also was pretty good at getting into trouble with the law – some minor issues, and some more serious.

"I was doing things that I shouldn't have been doing; thinking that I could get away with anything I wanted to do. The last time, it was for carrying a concealed weapon – not a gun, but a brass knuckles-type knife I had on me," Valencia said. "I went to court and the judge remembered seeing me before, and knew I'd been in trouble, so he gave me the option of going to the military or go to jail.

"Back in those days, that's what they did. There wasn't much to think about. I didn't want to go to jail."

Two weeks later, in December 1973, he enlisted at age 19. Four months after that, he was on his way to boot camp, still a tough, cocky teenager. It was a bit of a rude awakening when the bus arrived at its destination, but Valencia says he was up to the challenge.

"We were a bunch of young kids who all we knew about the military was what we saw on TV," he remembered. "We're all shooting craps and gambling on the bus, acting like we were big-time soldiers. Then when we got to Fort Ord, it was a shock.

"They started yelling at us, and screaming at us, getting us off the bus. It was late at night, and they gave us whatever basic stuff we needed, and we went to bed.

"At 4:30 in the morning, the drill sergeant came through with a trash can top – just like you see on TV – beating it and waking us up. Then we started in-processing, getting our hair cut, going to different classes, getting our uniforms. It was kind of fun, really. I kind of liked that discipline, that sense of order. A lot of people didn't like it, but I liked it right off the bat. I've always enjoyed a challenge."

After basic and advanced infantry training, it was off to jump school and Fort Bragg, where Valencia spent a remarkable 11 years in the famed 82nd Airborne, a fierce unit regarded as the best-trained light infantry division in the world, specializing in parachute assault operations, and tasked to "respond to crisis contingencies anywhere in the world within 18 hours."

He remembers particularly well the many jungle training missions in Panama, which were probably the toughest of many challenges faced by the young sergeant and his troops.

"I went to jungle school, like, five times," Valencia said. "That's tough training, because it's just a tough environment. Black palm, wild animals, moving at night … even the guys who were stationed in Panama thought we were a little crazy for moving at night in the jungle.

"You're walking at night, holding on to each other's gear; you've got zero visibility, because of the triple canopy. There's no moonlight, no nothing, coming in. It's like walking with your eyes closed, so you're holding on to each other and all of a sudden, the guy in front of you just – he's gone, off a 10- to 15-foot cliff."

Then, in October 1983 came a call to mount up, and this time it was more than a training assignment. There had been a truck bombing at a U.S. Marines barracks in Beirut that killed 220 Marines, 18 sailors, and three soldiers.

"That's where we thought we were headed," Valencia said. "But we went to Grenada, instead."

Along with Ranger battalions, Marines, Delta Force and Navy SEALs, the 82nd quickly overwhelmed the enemy forces in a battle that lasted only a few days.

"It wasn't a long battle. The Rangers jumped in, and then we were going to jump in behind them, but by the time we got there, they said, no, the Rangers have secured the air strip, so we're going to land – which was kind of disappointing, because I wanted a combat jump.

"It didn't last long, but we did get into some firefights. Something like that, you're not really scared, or nervous … pretty much focused, I guess you'd say. You've trained and prepared for it. We knew this was the real deal, and so we just got very serious and very focused. I told my guys, 'We're just going to do like we trained, and we'll be fine.' And we were."

He was awarded a Bronze Star for the Grenada mission, and went on to become a drill sergeant at Fort Dix, N.J., then an instructor at the NCO Academy at Fort Hood, then moved to operations NCO at III Corps headquarters, before retiring after a little over 20 years' service.

"My goal was to become a command sergeant major back in the 82nd Airborne, but when I called my infantry branch, they told me that I'd never see Fort Bragg again. They said I was a mechanized guy now, and there's more need for mechanized guys with my rank than there is at Fort Bragg. Which I understood, because there's only nine infantry battalions at Fort Bragg, so there's only nine sergeant majors.

"So I said, well, if that's not going to happen, I'll just retire. It was time to get out and pursue a civilian career."

Troy Summerill

Troy Summerill had ambitions of a short stint in the U.S. Army and then becoming a police officer, but after a few years of working as an MP, he decided his heart was not into arresting fellow soldiers and so he started considering other options.

"I'm too nice of a guy," Summerill, a Harker Heights resident, said, laughing.

"My job at Fort Sill was catching DUIs and DWIs. Sometimes, though, it was just better to take them home (instead of taking them to jail). Back then, Desert Storm was over and they were trying to get as many people out of the Army as possible – there was no leniency. You could be kicked out for anything.

"I'd see these E-7s and E-8s, young officers, who had made a mistake. It (drunk driving) is a bad thing, don't get me wrong, but people make mistakes. I make a mistake every ten minutes.

"It was a good career, and I was awarded and stuff like that, but it just wasn't for me. I worked with people who'd arrest their brother, and I thought, 'Man, that's tough. You can't just be a robot.'"

A native of Bremerton, Wash., Summerill graduated from high school in 1989 and decided to join the Army mostly for the education benefits. He served three years as a military police investigator, and got out as an E-4 (corporal) in 1992.

"(President) George Bush Sr. offered people the chance to get out early after Desert Storm (the beginning of ongoing U.S. involvement in the Middle East after the Iraqi invasion of neighboring Kuwait), and I raised my hand very high. I was ready to get out. I enjoyed my career, but I knew I was not going to be a career soldier.

"I just knew it wasn't going to be a career thing for me. I had other ambitions. I was going to come back home to Bremerton, which at that time was the second-best police department in the States. I actually knew people on the force, so it was like, 'Hey, I'm going to go to a really cool police department,' and then I got into hospitality and never looked back."

After leaving the Army, Summerill went back home to Bremerton, but his career path took a turn and he wound up working in the hospitality industry. That career eventually brought him to central Texas in 2009, when he worked several years as manager of the Shiloh Inn Suites Hotel in Killeen.

Unlike many former soldiers who settle in the Killeen area, Summerill was never stationed at Fort Hood, but now says there is no other place he would rather be.

"I used to run resorts, and our company was building one down here in Killeen. I was running one of their ocean-front resorts (Ocean Shores, Wash.), and they said, 'Do you want to take over the one in Killeen?'

"I said, no. That's what I said at first. But I wound up taking it, and it was probably the best decision ever. I love Texas. As soon as I got here, it was like, 'Wow, this is a pretty cool place.'

Now 50 years old and the father of six children, Summerill owns and operates three businesses in Harker Heights with his wife, Amy: POP CornUtopia, a gourmet popcorn shop opened in 2017; Sprinkles on Top Texas ice cream; and a mobile kettle corn maker.

Even though his stint in the military was short-lived, and his career in law enforcement did not materialize, Summerill says things seem to have a way of working out for the best. He enjoyed his time in the Army, is proud of his service, and has no regrets.

"None, whatsoever," he said. "It was great.

"My stepdad was a cop, so I thought that was what I wanted to become – a cop. I was headed in that direction, and then I got into hospitality and I liked the idea of being able to come home every night. So that kind of killed my law enforcement career.

"So far, everything is going well. Business has been pretty good. I'm not a millionaire, but I have no desire to be one. Just enough to pay the bills is fine with me. I'm still crossing my fingers … I've been very lucky."

Larry Letzer

Some say that finding an occupation you enjoy is a key to success and happiness, and longtime Copperas Cove photographer Larry Letzer is someone who certainly would agree with that idea.

Growing up in New Jersey as the son of a newspaper reporter, 82-year-old Letzer discovered his passion at an early age and thanks in part to the U.S. Army, was able to turn an early love for taking pictures into a successful, lifelong career that continues today.

"I was probably four or five, and my father took me to the Baltimore Sun (where he worked), and I went in the darkroom," Letzer said. "There was a very famous photographer in there – I don't remember his name anymore – and he showed me prints developing in the developer chemicals.

"I was blown away. I thought that was magic, and from that day on, I didn't want to do anything but be a photographer."

Letzer got his first camera, a Brownie Hawkeye, when he was seven years old and hasn't looked back since.

"I remember when I was nine or ten or eleven, I would go into my closet in my bedroom, put towels under the door to keep the light out, and develop film. Then go into the bathroom with a contact printer and print the pictures. I basically taught myself.

"I started doing photography as a young kid for the local newspaper. I was an apprentice starting out, and I rode with the photographer until I was old enough to get a driver's license.

"I worked for the chief photographer of Atlantic City when I was twelve. They had what they called the Atlantic City press bureau, and he was the chief photographer. Back in those days, they were doing all they could do to promote the community. I worked there and I also worked for the Atlantic City Press (newspaper).

"I lived a pretty charmed life back in those days. I was also a photographer for the Miss America pageant. I thought it was normal that every twelve-year-old kid got to meet Miss America, (astronaut) John Glenn, people like that."

Along with duties as water boy for the high school football team and scorekeeper for the baseball team, Letzer was a professional photographer all the way through high school. He did not have any interest in going to college – he now holds a bachelor's degree and master's degree in business – so he decided to join the military after graduating in 1956.

"I wanted to become an Army photographer. I didn't have the grades at that time (to go to college) or the inclination. I was in the Civil Air Patrol as a kid, and enjoyed the structure, so I thought I'd enjoy the Army and let them teach me photography."

His first hitch, Letzer served three years and then went back home to work for the Atlantic City Press. He went on to a stint with a commercial photographer in Atlantic City, and also became the official photographer for Steel Pier amusement park.

A few years later, Letzer decided that he might be better off going back into the service. This time, he stayed for 18 years before retiring from Fort Hood as a major the last day of October 1980 with 21 years' service.

"I missed the Army and so I re-enlisted in October 1963. Went back through basic (training) a second time; went to infantry training; and then went to OCS (officer candidate school) at Fort Benning.

"Back then, the Army did not want you to specialize. I think I would have made lieutenant colonel had I not been so vocal about staying in photography. I felt I could make a better contribution to the Army that way.

"The Army had a philosophy back in those days that every new second lieutenant would rise to the rank of major general. They wanted everybody to have aspirations of that lofty goal. I didn't have such high aspirations. I wanted to do photography. That's what I thought I was good at.

"I was a company commander a couple of times, and a battalion XO (executive officer), in non-photographic assignments, but most of my officer service – and all of my enlisted service – was in photography."

Letzer, a father of two and grandfather of two who has been married to Bernice for 58 years, came to Fort Hood in 1974 and settled in Copperas Cove, where he has lived ever since. That same year, he started Camera Artistry Photography, and grew the company into a thriving business that he has mostly turned over to his son, Larry Lee Letzer Jr.

"I started the photography business while I was still on active-duty. I'd take a months' leave in September and do the senior pictures at (Copperas) Cove High School. We've done the senior pictures in Cove since 1978.

"These days, I do a little bit of the finances, but my son does all the photography and everything else."

Looking back on his military career, Letzer, who served two tours in Vietnam (1965-66; 1969-70) says he thoroughly enjoyed being a soldier. He is not behind the camera nearly as much as used to be, but still enjoys riding his Honda Gold Wing three-wheeled motorcycle and using his commercial pilot's license to get behind the controls of an airplane now and then.

"Basically, boring holes in the clouds," he said.

"The Army taught me some discipline – not much, but some. It gave me a foundation. I wasn't ready for college when I was seventeen, eighteen years old.

"Like I said, I lived a charmed life. We've been fortunate in that we have managed to keep a payroll this entire time, even during COVID. I photographed lots and lots of celebrities when I worked at the Steel Pier; several Miss Americas. One day I was doing a convention, and Miss Universe and I walked from the convention hall to one of the hotels on the boardwalk, and I thought it was normal for every eighteen-, nineteen-, twenty-year-old kid to do that.

"The most remarkable person I ever photographed, I think, was John Glenn (legendary U.S. astronaut, and the first American to orbit the Earth, in 1962). That was a treat for me. He came to Atlantic City to speak, and I had to photograph him for the paper.

"I got there a little bit early, and he got there a little bit early, and we sat and talked. What a great guy."

Most people today have a camera at their fingertips, making everyone a potential photographer. To take better pictures, Letzer, who taught photography for several years at Central Texas College, offers this piece of advice:

"As far as having the eye (for composition), I think you either have it or you don't," he said. "My strongest suit in photography is getting people to relax. Making them feel comfortable with themselves, in order to get a decent picture of them.

"Enjoy your subjects. You need to be a people person to do photography – at least portraitures – so that you can make people relax."

John Cooney

For thrill-seekers and the curious who wonder what it would be like to fly one of the U.S. Army's powerful Apache tank-killing helicopters, former combat pilot and instructor John Cooney offers the following homemade simulation:

"When you're flying in an Apache at night, you're using FLIR (forward-looking infrared radar) to navigate, which is basically a camera that sees heat sources," said Cooney, a retired Chief Warrant Officer 4 and self-described Air Force brat who graduated from high school in Torrance, Calif., spent 20 years in the Army and 22 years as a civilian flight simulator instructor at Fort Hood, and now lives along the southern shores of Belton Lake.

"You've seen the movies, and they show these really crisp pictures that look like a normal photo, but everything is in green? Well, it doesn't quite work that way.

"Especially in the early Apaches, it was more like a photo-interpretation of what you were looking at, rather than seeing crisp, beautiful pictures. And our field of view was less than 40 degrees, so you're expected to fly a 15,000-pound helicopter, at two in the morning, using a not-so-great FLIR picture – and you're supposed to operate at 120 miles an hour, fifteen feet above the trees, going out to look for bad guys.

"If you ever want to get an idea of what it's like to fly an Apache, go get a cardboard toilet paper tube and a black plastic bag. Punch a hole in the bag and push the cardboard tube through the hole. Go get yourself a piece of green cellophane wrap, and tape that to the outside portion of that toilet paper tube. Put the bag over your head; hold the tube to your right eye; and get in your car and drive around town. Or just walk around the house.

"That'll give you an idea of what it's like to fly an Apache."

Cooney was born in 1950 at the University of Iowa hospital. After a childhood filled with moving from place to place every few years, he graduated high school in 1968 and went to college for a year. Unfortunately, he spent more time concentrating on spinning records as a disc jockey at the college radio station that he did on cracking the books.

"I didn't do too well," he said, "so I lost my deferment.

"We had lived in Okinawa for a couple of years, and while we were over there, the Army recruiter had come in on one of those recruiter days, and he showed a small video of warrant officer flight school – and that had stuck in my brain. I had wanted to be an Air Force jet pilot, but you have to have a college degree to get that, and since I wasn't going to make it through college and I still wanted to fly, I chose the Army route.

"Went down and took all the tests and passed them with flying colors. I got sworn in and everything, and then I told my parents.

"They were not happy. Dad wanted me to go into the Air Force, and my mom didn't want me to fly helicopters because this was kind of the middle of Vietnam and helicopters were getting shot down pretty regularly, and she just knew I was going to get killed.

"Dad would bring people over for dinner who had just gotten back from Vietnam, or who had flown missions over there. They told me, yeah, the average lifespan of a helicopter pilot is a month. I said, 'I don't care – I'm going.'"

Cooney reported to basic training at Fort Polk, La., in September 1969, went to flight school, got married, and by the time he turned 21, was behind the controls of an Army attack helicopter searching the jungles of Vietnam for enemy troop positions.

"I was stationed with the 1st Cavalry Division and flew with the same unit for the year I was there," he said. "I got 880 combat hours, and I was a very lucky person in that the whole time I flew, my aircraft never took a bullet. Never got hit. My scout would tell me, 'Hey, you're getting shot at.' And occasionally you could see a bullet – a tracer – go by, but never took a hit."

While he escaped being on the direct receiving end of enemy gunfire, Cooney handed out plenty of pain during his scouting missions. One of those times came when a jungle run uncovered a "rest and relaxation" camp for the North Vietnamese and Vietcong.

"We had an aircraft – an OH-6 light observation helicopter we called the "Loach" – and we would take one of those and one Cobra. We were called the pink team, because the gun platoon is the red team. Scout platoon is the white team, and if you had infantry support, they were called the blues.

"So, if you had one Cobra and one (OH-)6, then you had one red and one white – the pink team.

"We normally didn't fly at night. It was almost dark, and usually our OH-6 would fly up with us and we'd fly home. The whole group of us – maybe five, six, seven Cobras and the same number of 6s, and a couple of Hueys – would fly home.

"My scout loved to fly down on the trees, so he was a thousand feet below me and just in front of me. All of a sudden, he starts climbing like a bat out of hell, screaming, 'Taking fire; taking fire.'

"I put the nose over and shot all of my rockets into that area. I pulled up next to him and he read me some coordinates, and I checked it and we agreed on where this had happened. I called back to the TOC (tactical operations center) and they said, 'We've never taken fire from that area. What's going on?'

"For three days, the Air Force bombed the holy snot out of it. We went in there and shot thousands of rockets into the area, and every day somebody took fire – so we just kept hammering it.

"One time, I went in and my scout ran in and said, 'I found this huge bunker,' and threw out a smoke grenade. I called the guys who were stacked up above me – two more Cobras outfitted with nothing but rockets; an Air Force spotter plane; a set of South Vietnamese Air Force A-37s with bombs; and a pair of F-4s (fighter-bombers).

"So, my scout runs in, finds this bunker, throws his smoke, and just as he's getting ready to run away, a guy – one guy – jumps out of the front of the bunker, raises his gun and starts shooting. We see the tracers going up, so my front-seater puts the mini-gun on it and he squeezes the trigger. He's shooting two thousand rounds a minute back at this guy. I put the nose into a dive, (and) shoot a bunch of my rockets.

"As I'm breaking away, I hear this thunderous roar go by me, and when the dust settled a half-hour later, there was nothing but a huge smoking hole where that bunker had been."

Cooney continued flying Cobras and Apaches throughout his career, served as an instructor pilot, retired in 1989 and went to work as a civilian flight simulator instructor at Fort Hood. He also worked for a while as operations officer for the post's Training Support Center, and as a lead contract officer representative for the Department of Public Works.

He retired for good on Dec. 31, 2019 with a total of 50 years' government service and 6,000 hours flight time, including 4,500 hours flying Cobras and 1,000 hours in Apaches.

Now 70, the father of three sons and six grandsons who was a founding member of the Killeen Evening Rotary Club and served twice as club president enjoys playing golf and riding his motorcycle when he gets the chance. He prefers long-distance rides, like the day a few years ago when he logged 1,207 miles from Harker Heights to near Rapid City, S.D.

"We (he and wife of 33 years, Patsy) had several things planned, but COVID stopped us from doing a lot of traveling.

"I enjoyed my military career. Nowadays, people tend to say 'thank you for your service' a lot. I'm thankful for that, because when you came back from Vietnam, most people didn't want to talk to you at all. They'd spit at you and called you baby-killer and all that.

"I didn't get spit on, but I did get called some names.

"So, when people say thank you for your service, my answer is, 'It was my honor and privilege to be able to serve my country.' That's the way I feel. I was sworn in to do a job, and I will do that job to the best of my ability until I die."

Raven Garcia

When Copperas Cove resident Raven Garcia decided to join the military after high school, she was like many other young recruits looking for a chance to get away from home, see new places, and maybe find a meaningful place in the world.

Little did she know that her short stint in the U.S. Navy would soon become life changing.

"The main motivation for me was, I wanted to get the hell out of Dodge," said Garcia, who was working at a McDonald's restaurant in her native Corpus Christi, alongside the Gulf of Mexico about 280 miles south of Killeen-Fort Hood. "I wanted to figure out who I was; I wanted to see the world; have a steady paycheck.

"I joined the delayed entry program, and you could take up to a year to actually leave for boot camp. That was OK with me, because I knew I was going to do something besides flipping burgers, which is exactly what I was doing at the time. My father – who has passed – was also a sailor. He was in the Navy for three years. And my stepdad, Loren D. Lee, who was a Navy radioman and part of a flight crew serving on a blimp, was a huge inspiration and motivator for me.

"I didn't want to go to college. I thought I'd join the Navy for a few years, and who knows? Maybe I'll like it and it'll be a career. I was just sick of school."

After graduating in 1986 from Richard King High School, Garcia hung up her McDonald's uniform and headed to basic training at the Navy Training Center in Orlando, Fla.

Boot camp was surprisingly easy, she says, but what came next was more than a little disappointing.

"It was actually kind of a joke," Garcia said. "I thought it was going to be like the Army – running obstacle courses, getting in the dirt, jumping in foxholes, playing war games – but it was nothing like that. The most that we did was run around on the blacktop at a pace you could walk. We did pushups. But mostly we practiced marching.

"Boot camp in the Navy is more like school than anything else. It's not like the Army, where you learn to shoot guns (and) hand-to-hand combat. It's more about schooling. I had to drop twenty pounds to get into the Navy, and I gained it back during boot camp. That's how lame the physical fitness part of it was.

"I went in non-rated, meaning I didn't have a specific job that I signed up for. I wanted to be a photographer's mate. All through high school, I was really into photography. I told the recruiter what I wanted, and he said, 'If that's what you want to do, you can pick that school up after you get out of boot camp.'

"I believed him. He lied to me. I didn't know any better."

After a month of being in limbo, Garcia was assigned to an overseas duty station in Sicily. She was not happy.

"I was scared s—tless," she said. "I wanted to travel, and I wanted to be in a new world, but I was hoping I could just get used to being in the military first … before they sent me across the globe.

"But it turned out to be a really positive experience. A lot of great things happened. I made some really great friends, and I finally got to shoot pictures. I didn't get to go to the school that I wanted to, but I had my camera and I came across a flyer stuck on the billboard in the chow hall that said the public affairs office was looking for volunteers to help shoot pictures for the paper, and to write articles. So, I volunteered.

"They started sending me out on assignments, and at the time, I was a file clerk for the judge advocate general's office. So, most of the time, I would run legal documents back and forth to the executive officer and commanding officer, for them to sign.

"If I was having to do different errands and I was going to be in an area where there was an assignment to shoot, I would take my camera with me and shoot the assignment. Kind of kill two birds with one stone. I used to get in trouble all the time.

"They would say, 'Airman Garcia, that camera is not part of your uniform, blah, blah, blah.' I would tell them, 'I know, but I've got to shoot this assignment for the paper.' I did that a lot, and they would fuss at me, but they admired that I had that kind of drive and motivation.

"One afternoon, I was out running errands and when I came back to the office, everybody was there. They were all standing there looking at me when I walked in, and I was like, 'Oh, I'm in trouble for something.'

"They all had this straight face, and then they all started smiling and laughing, saying, 'Congratulations!' They had a cake that said, 'Congratulations, Airman Garcia,' and they showed me the paperwork that officially attached me to the public affairs office."

Her dream assignment proved short-lived, however, when Garcia made a monumental decision to confide a career-threatening secret to her supervisor.

She confessed to being homosexual.

"At that time, it was not OK to be gay and be in the military. They were actively doing what we called 'witch hunts,' where if they heard a rumor that a certain sailor was gay, they would do undercover stings and b.s. like that to catch people and kick them out.

"They'd give a dishonorable discharge, even if you were a (squared-away) 4 0 Sailor. If you were caught and there was evidence, they would just kick you out.

"I met this chick and I thought I was in love with her. I really thought, 'This is it.' I thought we were going to have a life together. She got reassigned back to the States, and I wanted to go with her, so I turned myself in.

"As a JAG clerk, I had access to the Navy law books, and so I researched what I was getting myself into. I wanted to be prepared mentally for whatever I was going to have to go through.

"I went to the chaplain first and turned myself into him. But the process wasn't happening fast enough. They weren't doing anything. They wanted to send me to Spain for a psych evaluation, and I was like, 'Ok, the sooner, the better. Whatever it's going to take.' But they never did; they kept dragging their feet. And that's when I decided I'm just going to incriminate myself, and whatever happens, happens.

"I wrote the executive officer a letter. He called me into his office and asked me, 'Why are you doing this? Has anybody threatened your life?'

"I said, no. I just didn't think I could be a 4 0 Sailor while I was hiding who I am. It's just impossible. I thought I'd wind up getting kicked out anyway, and I'd rather just bow out now before I get too deep into a career.

"I couldn't really be who I was. I would always have to be hiding. I lived my entire childhood that way. I was hated by the church. I was hated by my peers. I just didn't want to torture myself that way anymore. I wanted to be free – free to be me.

"I was really fortunate. I worked closely with the executive officer and the commanding officer, and they got to know me more on a personal level, and they made the decision to give me an honorable discharge."

After her release, Garcia headed to Florida to pursue her new relationship, but things quickly fell apart. She stayed in Florida, working various jobs, and eventually becoming a successful self-employed massage therapist, a career she still enjoys today. Eventually moving back to Texas, she discovered a love for working with metal and now has added metal artist to her resume, crafting things like sculptures and custom furniture pieces.

The 53-year-old was only in the military for a year-and-a-half, but it was an important chapter in her life, and one she looks back on with pride and satisfaction.

"I think I would have had a great military career," she said. "I really had myself in a great position to advance pretty quickly, and to get the school that I wanted.

"It was definitely a good thing for me. I'm a firm believer that your attitude has a lot to do with anything that you do. I was very open-minded about my experience, and I just decided that no matter what, I'm going to enjoy the journey. I'm going to learn; I'm going to figure out who I am; and it's going to be a good thing – and it was.

"I learned a lot about myself, and I met some really great, inspiring people. Got to see a little bit of the world. Drank some great beer.

"Being gay and joining the military today is not an issue, unless you make it one. Gay people who take their military career seriously make that the forefront of who they are and how they operate. Being gay is like the color of their hair; the color of their skin; the color of their eyes. It's a part of your DNA, but it's not like you walk around going, 'Back up, I'm gay.'

"Gay people are just like everybody else. They just happen to love somebody different."

Robert Crusing

Retired U.S. Navy Master Chief Robert Crusing was never on the front lines during his tour in Vietnam, but an incident late one night in Saigon apparently left lasting emotional scars that are sometimes still painful today.

A Chicago-area native who lives now between Kempner and Lampasas, just west of Killeen-Fort Hood, Crusing graduated high school in 1966 – the height of the Vietnam War – and went to college for a year before Uncle Sam came calling.

"They decided not to invite me back for a second year of college," Crusing said. "The campus I was on was set up so that you had to walk about a quarter of a mile to your first class, and at about an eighth of a mile was the gym. So, you know …

"This was when they had (draft) numbers, and I had a low number. I didn't want to be a ground-pounder, so I went ahead and joined the Air Force. I was in the Air Force for four years, including a year in Vietnam (Saigon).

"Then I got out and went back to college. This time, I was able to pass, and I got two years in junior college, then I went to the University of Chicago. I had an 'A' average when I went there, but they were a little tougher than junior college. I took a third-year chemistry class and there was a hundred and four people in the class. I'm going, 'Oh, my God. How am I going to do this?'

"I passed, but I had like a 'C' average, and I was afraid that was going to ruin any possibilities of getting a scholarship for grad school, so I looked around and tried to get back into the Air Force. They said, 'Oh, sorry.' If I wanted to come back in, they wanted me to start all over from day one. I didn't understand that. I was an E-5 when I got out.

"Right across the hall there was this Navy guy going, 'Hey, do I have a deal for you.' So I ended up joining the Navy. They said I was going to start as an E-4. Well, they lied, and I started as an E-3, but I didn't have to go through basic training again. So that was a good thing.

"It all worked out. Eighteen years later, I retired as an E-9 with 24 years of service overall."

It was during his Oct. 1969-70 tour as a fire control technician with the Air Force in Vietnam that Crusing suffered what doctors later determined was some sort of emotional trauma that affected him throughout his military career and beyond.

He was relatively safe and sound living and working every day inside Tan Son Nhut Air Base in Saigon, reading and interpreting aerial photography images, and not out humping through the jungle, dodging bullets, bombs and land mines. But then alarms sounded one night and nearby explosions started rocking the compound.

"I was in a new, well-constructed, air-conditioned, window-free, brick building. For being in Vietnam, it was outstanding duty," Crusing said. "The armory was right there. You couldn't ask for a whole lot more than that.

"One night, in the middle of the night, we got some incoming. Sirens and alarms went off. My barracks, they had sand piled up four feet high around the building, and there was no place for me to go. So, I just pulled my flak jacket up over my head and went back to sleep.

"I didn't think much about it. When the rockets were hitting, there was nothing else for me to do, so I just pulled the flak jacket over my head. They said normal people don't do that, in that kind of situation.

"There was nothing else to do. There was no place else to go. The rockets were falling 60, 80 feet away. Where was I going to run to? So I just put the flak jacket over my head and went to sleep. I don't know of anybody else in that barracks that did anything differently.

"After that, I started having some problems. I had some episodes where I couldn't handle a lot of things that dealt with the passing of military (personnel). You know those missing man flyovers? I'd get all teared up over those. I was stationed in the Navy in Hawaii, and the stern of our ship was right across the harbor from the (USS) Arizona Memorial. I couldn't even look at it. I would just get choked up. I was there three years and never went to the memorial.

"The Vietnam traveling wall came to my hometown. I got about 30 feet away and I was crying. I thought, 'You know, there's got to be something wrong with me.'"

Eventually, Crusing saw a counselor, was diagnosed with clinical depression, and prescribed medication to help. He retired from the military in January 1994 at age 46, and immediately went to work as a civilian fire control instructor – the same thing he did in the service.

Now 72 years old, Crusing moved to central Texas three years ago to get away from the frigid winters up north, and to be near his daughter, who lives in Harker Heights. Married to wife, Linda, for 34 years, the couple has four children.

Crusing says he is doing well these days. He is proud of his military service and looks back on those years with pride and satisfaction.

"I've had a blessed life," he said. "I made E-5 in the Air Force in three years. I made E-9 in the Navy in sixteen years.

"I think I was a positive influence on my younger petty officers and recruits. I tried to present a positive picture when I was teaching, even when I got out of the military. I enjoyed every minute I was in. I've been very fortunate.

"I have done a lot of things – and always ended up on dry land."

Cindy Aswell

Cynthia Aswell worked as a bartender after she graduated high school, living without much direction or purpose, when she made a decision more than 20 years ago that changed her life.

She joined the military.

"After high school, I kind of got confused about what I wanted to do, and I think I just took the easy way out," Aswell said. "I was just hanging out with my friends – no college; no career plans; no drive to do anything, really. I'm not a big drinker and I don't do drugs, so it wasn't anything like that.

"I had kind of gotten to a point in my life where I had no motivation to do anything or go anywhere, and to just buck the system at every turn. At some point, I just decided I needed somebody to give me a swift kick in the rear end. Once I made the choice to join the Army – and committed myself to it – it changed the very essence of who I was. They taught me to be a better person; they showed me a better way."

Aswell, a mother of four who lives near Fort Hood, was born and raised with four brothers and a sister in Marietta, Ga. She graduated from high school in 1992 and joined the U.S. Army when she was 22 years old.

She reported to boot camp at Fort Jackson, S.C., and promptly experienced a bit of a rude awakening.

"I went from doing whatever I wanted to do – nobody telling me what to do – to somebody right there in my face. You know, the round brown tapping your forehead is enough to disconcert anyone."

Training was tough, but she never considered throwing in the towel.

"From what I understand, it was a lot harder back then than it is now," Aswell, 46, said. "The dynamics were a little different. Not quite the, uh, kinder, gentler Army.

"I think the biggest shocker is that there's no family – no one to turn to and say, 'Hey, look at me; feel sorry for me; talk to me.' None of that. You literally learn to rely on yourself and the soldiers around you.

"It definitely had its moments – I'm not gonna lie. But I took it as a challenge. If you think I can't do these things, I'm going to show you how I can. I have this thing about me that once I start something, I won't allow myself to quit. I've been that way my whole life. If I commit to something, it's over. I'm going to do it. In my career, that's what I tried to teach any female soldier I had – it's not that you're a female; it's that you're a soldier; and this is how you do it."

Aswell served a total of 12 years in the Army, working in flight operations (maintaining records and overseeing aircraft rescue missions) at various duty stations, including Fort Campbell, Ky., Fort Rucker, Ala., and Fort Hood. She was deployed to Iraq from 2007-09, a time she remembers as extremely challenging but also rewarding.

"It had its moments that were difficult, but I would never change a single moment from what I learned, and how it shaped me into a better person today.

"It challenged you. It showed you that there are going to be things out there that are not going to be pleasant, and that you can deal with them and overcome them. It teaches you to have a certain amount of reserve and gives you that drive to go through those difficult situations and still come out a normal person on the other side.

"And when you have soldiers underneath you, they expect you to do that."

Aswell achieved the rank of E-5 (sergeant) before she left the Army in 2009. She was married then and was faced with the decision of staying in the military or keeping her family together.

"At that time, my ex-husband had given me an ultimatum. He said, 'It's the military or your family.' My family was very important to me, and I wanted to give them everything they needed. So, I decided to step away.

"It was the hardest choice I ever made, and one I continue to regret. After that, it was a series of trial and error … I tried to be a dental assistant, and I tried being a medical assistant. Dental assistant was a very short stint, and I loved every second of being a medical assistant, but the unfortunate part of it was working 12-hour, 14-hour days.

"It took me away from home so much, and the kids were getting frustrated with it. They wanted me to be able to spend some time with them again, so I stepped down. But I was going stir-crazy. I couldn't just sit around, and that's when I found the Kempner VFW."

Two years ago, Aswell began working as adjutant and service officer for VFW Post 3393 in Kempner, a position she holds today and one that she thoroughly enjoys. She is passionate about helping fellow veterans, and since the start of the COVID-19 pandemic earlier this year, has become involved in a concentrated effort to save VFW posts, which have experienced serious financial hardship due to restrictions placed on their bar business, a primary source of revenue.

"People are starting to see there's a problem," Aswell said. "The veterans deserve the help.

"In a local area, one of the first things the VFW provides is an outlet for them to come in and be able to be around other veterans who understand what they go through on a daily basis – whether it be their body aches or their mind aches. They have that ability to communicate, and hopefully that helps prevent people from going down a path that is dark and not good.

"Secondly, we show you things you can do to integrate yourself into the VFW, so that it gives you a new purpose, and you are able to then step out and start helping others. We also provide things like rental assistance, rides to medical appointments, and we do community service.

"I think the biggest thing for me is I just want people to understand that the VFW is not age-based. Unfortunately, a lot of people think of the VFW as Vietnam veterans. In reality, the VFW is an ever-evolving entity. They are always moving forward and opening their doors – and their arms – to anyone.

"It doesn't matter who you are, and it certainly doesn't matter your age. Whether you're 22 or 65, it doesn't matter. We have active duty (soldiers) who are members now. We encourage every single one of them to join. Everyone is welcome.

"The VFW came around at a point that I was lost, and without hesitation, they opened their arms and their hearts to me," Aswell said. "When something like that happens, you tend to have a certain amount of passion towards that. You helped me during my difficult times; now it's time for me to turn around and help someone else.

"There are some amazing people there who are so good at guidance and motivation, and just giving you that camaraderie you miss when you leave (the service). I finally found a group that understands that dynamic.

"For the community itself? Just remember that they're there. They're there for the community every single day – the veterans and the civilians. The VFW is there to support everyone."

Chris Owusu

When Killeen resident Chris Owusu graduated from high school 20 years ago in Alexandria, Va., his plan for the future was fairly simple: earn a spot as a walk-on for the basketball team at nearby Division II Bluefield College.

One of four boys in a family headed by parents who immigrated to the U.S. from Ghana, Africa, Owusu was a talented but quick-tempered young man and soon found himself in serious trouble at school for fighting. In fact, he got kicked out of college pretty much before his career began.

He decided to try again, enrolling this time at Northern Virginia Community College, but academic issues caught up with him and he found himself on the outside looking in, trying to come up with a Plan C.

"At that age, I really didn't have anything in mind, as far as future endeavors," Owusu said. "I didn't know my 'why.' I didn't have the motivation, and I was just not focused.

"At that point, I'm still a man of faith, so I'm praying and I'm thinking, 'OK, I have to capitalize on my strength, which is to overcome adversity.' Our house was by a military base – Fort Belvoir – and my friend's dad was military. I would see the military, and so I always knew that was an option. I was physically fit, and so I knew the military would be a good choice for me."

Owusu was 22 years old when he went to see a U.S. Army recruiter. He was not sure what he wanted to do in the military, but the recruiter helped him pick a job that wound up being a good fit for him.

"My recruiter was very upfront, and to this day I'm thankful for that," he said. "I wanted to do something to be a part of history – I didn't want to be behind a desk in an air-conditioned office. So, I chose cavalry scout."

From there, it was on to boot camp and advanced individual training at Fort Knox, Ky. He was the first in his family to be a soldier, but the rigors of basic training that unnerve some fresh-faced recruits were no big deal for Owusu.

"Because of my upbringing, boot camp was relatively easy," he said. "The yelling, the screaming, the pseudo-intimidation … all that didn't move me. My parents had already taught me that you have to submit to somebody in order to grow. So that was already in my mind. I learned quickly to stay under the radar."

Sixteen weeks later, Owusu arrived at Fort Hood, his first duty station, where he became a member of the 4th Infantry Division. A year later, he was on his way to Iraq, for the first of two combat deployments.

He arrived the first time in October 2005, conducting reconnaissance missions and surveillance duty at observation posts.

That December, he was part of a unit sweeping a village when tragedy struck, and two of his buddies died in land mine explosions.

"It was December 27th," he said. "They were very close together. One (blast) went off, and that triggered the other one. My back was turned, so I didn't see anything – I just heard the explosion.

"There was a lot of chaos and confusion, and then once you came back to the vehicle, you are looking for the people who are with you. I'm asking for the one guy, Josh, who was the driver – 'Where's Josh at?' And everybody just looked at me.

"I could curse God for things like that, but I realize that would only hurt me.

"Actually, it helped solidify my faith. It showed me that you have to relish every day, because every day could literally be your last. You never know what's going to happen.

"I look at it like their life was a gift. God gives and God can take away at any moment. I'm not the person who can determine how or when somebody leaves. I can just be grateful that I had the opportunity to be around them, and maybe bring God into their life.

"I understand that this is his Earth, and I'm just grateful to be on it."

After that first year-long deployment, Owusu returned to Fort Hood and then shipped out again to Iraq from September 2007 to November 2008. He left the Army a year later, one day after a U.S.

Army major opened fire on the post, killing 13 and wounding more than 30 others.

"I was on post that day," Owusu said. "I had a little boy I was mentoring, and I was going to go watch him skateboard. That's when the shooting started. I went to the visitor's center, and they locked down post until like 9 p.m."

Like so many soldiers, Owusu remembers exactly how long he served, almost to the minute. He never planned on making it a career, but says he is "extremely grateful" for that decision he made two decades ago to enlist.

"I was in for five years, six months, and three weeks," said Owusu, who recently celebrated his 39th birthday. "I knew going in that I wanted to join the military to supplement my future. At that time, I had gotten into a fight in high school, got into a fight at college, and so I already had a minor record. I knew that I had to have something to supplement my resume.

"I knew I was going to use my GI bill to get my bachelor's degree (in criminal justice) and go into education. I knew that because what I enjoyed most about the Army was helping the Iraqi children – so I kind of put two and two together. It was very minor, but we used to give them crayons, book bags, coloring books, and even though it was something small, I found a lot of fulfillment in helping them, so I knew that was something I wanted to do with the rest of my life."

Now, Owusu is in his sixth year as a public school teacher. He still enjoys playing basketball, is active in his church, and a few months ago, he helped found 'Boys Matter to Men,' a youth mentorship program in Killeen.

"We have heavily screened volunteers, and what they do is they interact with a young man, and they instill certain values based on our program. A young child is more moldable than a stubborn adult. It's going very well. We actually have a waiting list right now.

"I'm definitely grateful I went in the military – 100 percent. One thing I really miss is the camaraderie with my fellow soldiers, but I'm still in touch with a lot of my buddies. You can't replace those bonds you form."

Harald Weldon

Harald Weldon had plenty of close calls as a U.S. Army helicopter pilot in Vietnam, but one night in particular stands out more than most.

"Some missions were a lot more intense than others," said Weldon, who flew mostly UH-1 Hueys while logging 47 combat missions during his two tours from 1968-70. "Being in aviation was having extended periods of boredom punctuated by short periods of mass hysteria.

"Things happen so quickly. You do have spurts of fear during a firefight, or during a mission where you're shooting and being shot at, but you're so focused on doing your job that you don't really get scared – until after you get back.

"I actually crashed twice. One time, we were going into a landing zone, in the jungle, and the tail rotor caught a tree.

"That was kind of spooky, because they couldn't get us out and we had to stay there overnight. Aviation guys don't like to do that. We weren't familiar with that area, and we didn't have (a lot of) weaponry. I don't think I even had my M-16 at the time – I just had my .38 pistol. You're kind of in an enclosed area, as far as the jungle is concerned. You don't know … if you had to go out, which way would you go?

"I don't know if we were actually scared – we were more or less just on guard. We were young and invincible, you know, and we had other troops protecting us. The infantry troops are firing all night long, keeping the enemy at bay.

"They came in and dropped rigging materials – straps and stuff like that – so we could prepare the aircraft and lift it out. After they lifted it out, there was enough room in there for them to bring in another Huey and take us out."

Weldon spent the first eight years of his life in Germany, where he was adopted by an American soldier stationed there and his German wife and brought to the United States in 1954. The family was stationed at Fort Sill, Okla., then transferred to Albuquerque, N.M., where young Weldon spent his formative years and graduated from high school.

After graduation, he and a buddy decided to join the military.

"I didn't know what else I wanted to do," Weldon said, smiling. "This was 1964, and it was kind of a trend back in those days. It was not that unusual to join the military at that time."

Before long, the 19-year-old was off to boot camp at Fort Polk, La., and like most old soldiers, memories of those first few days are still fresh and easy to describe.

"Oh, heck, yeah. It was different than it is today. A lot of yelling and screaming. Short, short haircuts – that was one of the first things that happened.

"There's so much going on that you don't have much time to think about it. Back then, it was one thing after another. And you're not the only one in the game. Everybody else is going through the same situation. My dad was in the military, so I kind of knew what to expect. Not to mention the fact that dad was kind of a taskmaster himself. So it really didn't surprise me that much."

After basic and then advanced training, Weldon was assigned to field artillery duty in Germany, which was uneventful, and led him to make what proved to be an important decision.

"I didn't see much of a future in field artillery, and back then, after you'd been in for a year, you could re-enlist and get the option of a different career path. So, I did that, and I went aviation.

"I was sent to Fort Eustis (Virginia) to train on helicopters, and by then, Vietnam had started, so I knew what my next duty assignment was going to be."

After meeting his future wife, Peggy, followed by a stint at Fort Hood, it was off to war.

"I took her back to Virginia, then I came back here," Weldon recalls. "We went on a train to California, and there, we got on a ship. There were actually three ships. I was on the (USNS) Upshur, and believe it or not, the Upshur was the same ship that my parents and I came to the United States on in 1954.

"It took 30 days to get from California to Vietnam. It's kind of boring, but you stay kind of busy, too. You're doing a lot of busy work, and once you got way out to sea, you practice firing your weapons – firing machine guns off the ship. You did PT (physical training) on the ship.

"The only time we stopped was at Subic Bay in the Philippines for 24 hours. You had to get off the ship and … stretch your legs a little bit. The Air Force had a base there, so we went to the clubs. Hey, you're on your way to Vietnam, so you had an excuse to act out a little bit."

Now 73, Weldon retired from the military at Fort Hood in 1986 as a first sergeant, and the recipient of two Purple Hearts, a Bronze Star, several Air Medals, and other decorations.

The father of two went on to a second career working another 26 years in quality control management for DynCorp and Raytheon. Now, he likes to tinker with vintage automobiles, spend time with his five grandchildren, and keep tabs on his old Army buddies.

For Veterans Day, it is sort of a tradition for Weldon to take advantage of the holiday specials at the Olive Garden restaurant in Killeen, but this year, he might not be able to make it.

"Oh, sure, it's a big deal because all my friends are veterans, and we look forward to the free lunch," he said, smiling. "I may not be able to go this time, though. I'm on a diet.

"I went to lunch today with an old sergeant major friend of mine, and we went to McAlister's. I studied that menu for an hour, trying to find something that would meet my (calorie) limit. Finally settled on soup.

"No. 1, I'm too cheap to buy new clothes, and no. 2, I like playing with my toys out there in the garage – cars and stuff – and if you get too big, you just can't crawl underneath there."

Pamela Grant-Rowland

Pamela Grant-Rowland was born in Dallas, attended elementary school in Killeen and spent a number of childhood years in Europe and Central America before eventually following in the footsteps of several family members and joining the military.

The now 53-year-old small business owner and divorced mother of three wanted not only to serve her country, but also to finish her college education and build a better life for her family.

"I've been a military brat, then a military dependent, and then I joined myself," said Rowland, president of AmeriVet in Copperas Cove, a company that manages renovations to aging government buildings on nearby Fort Hood.

"My great-grandfather was half-Indian and served in World War II as a bartender in Hawaii for the Marines. My grandfather went at the same time, but he was in the Navy. My stepdad served; my mom served; my ex-husband served; and I served.

"My whole family wanted to serve. It's part of a tradition. I think all people should try it, so they can understand more what serving your country is all about."

After graduating from high school in Kentucky, Rowland married her high school sweetheart, a U.S. Army soldier, and the couple started building a family. Three kids later, they were divorced and she was enlisting in the U.S. Army.

"Once the kids were older, I wanted to finish school – in between the first two kids, I was in college – but I couldn't do it the way I wanted to because of money.

"So, I worked it out with my mom. If she would take the kids, because she was retired, I would join the Army. My husband didn't like the idea. He wanted me to be a homemaker and he also wanted me to work – and I couldn't do both. So, we wound up getting a divorce.

"The kids lived with my mom in Kentucky, until I got my E-5 (sergeant). Then, the kids were with me and we traveled to New York and Hawaii. I wound up falling in love again in New York, and we got married in Hawaii.

"To make a long story short, he was retiring after Hawaii, and he didn't want me to stay in. I was doing very well. I was a flight medic, and I was recommended for PA (physician's assistant) school, and I was also trying to finish my degree (in business marketing and advertising). He wanted to retire in New York, and I wanted to come back to Texas, so that was that."

The couple went their separate ways and Rowland left the service not long after a pair of horrific tragedies struck while she was stationed at Hawaii's Schofield Barracks in 2001. One disaster was up close and personal, while the other happened thousands of miles away.

"I've not been to war, but I've seen what war is," Rowland said.

She was participating in a training exercise dubbed "Lightning Thrust Warrior" when a pair of Black Hawk helicopters collided at the Kahuku Military Training Reservation on Oahu's North Shore. Six Schofield Barracks soldiers on one of the choppers died and four more were injured.

Rowland was one of those assigned to assist at the crash site.

"I was required to go up there and figure out who all the people were," she said. "It was pitch-black dark out in the middle of nowhere, on hills. It's muddy because it rains up there every day.

"I was not on that aircraft, but somebody who was on it was on one of my aircrafts the day before and got me in trouble when he walked toward the rotor blades from the front. You never walk to a helicopter from the front while it is running because they pitch forward. He almost had his head cut off, and that's exactly the way he looked when I saw him dead.

"I tried to crawl in there and check for people still alive. I had no idea I was risking my life. It was dark and I didn't realize that thing was sitting so close to the edge of a cliff.

"It was absolutely terrible. When I flew with the dead back to the hospital, it was me and a general, some dead people in the back. The doctor that received them started crying, and he said, 'I don't know how you did it.'

"I shut down for a long time after that. Some of the people I worked with had experience with those kinds of things, but I had never seen anything like that before."

About six months later, terrorists flew commercial jet airliners into the World Trade Center towers in New York City.

"I was sitting in a surgical tent at the airfield, and I thought, 'Wow, does this ever stop?' We were all locked down, and I dealt with how we shut down the entire island of Hawaii. No planes were in the air; no food was coming in. The island really started to get depleted.

"The way we saw all the empty shelves here (earlier this year) during COVID, it was worse than that.

"Then they opened everything back up and we went on. After all of that, my husband at the time said, 'You know, the kids need to have a mom.' I said, yes, let me finish my term. So, I was in for six years. I was 31, 32, and E-5 promotable when I got out (in 2002).

"I wanted to come back to Texas and live near Fort Hood because this is where I went to (Peebles) elementary school when my mom was stationed here, and we lived in Killeen. She was in the Women's Army Corps and worked for (former III Corps and Fort Hood commander) Gen. (Robert) Shoemaker."

For the past 11 years, Rowland, who is also a talented artist, has been running her company and becoming increasingly involved in the local community through various volunteer efforts. She is a spiritual seeker and believes her experiences in the Army set the stage for the rest of her life.

"I think my soul needed to experience what I went through in the military," she said. "I think those people that died … I hope that they didn't suffer. I talked to their families, and I think and I hope that I helped them.

"I went to the wake. It was at a beautiful beach park – remember, it was in Hawaii. There was food and music and everyone was hugging each other, trying to celebrate life. So I learned about celebrating life.

"I wouldn't change anything. That incident taught me how to cope, and it gave me an awareness of how some people act under pressure, (and) how I acted under pressure. It made me a stronger person, and if something like that were to happen today, I would go into automatic help mode. I would do all the right things.

"Do you always notice the helicopters when they go over? I do. Every single time. Sometimes it wakes me up in the middle of the night. When I hear the local train, I go back to sleep. But when I hear the helicopters, I perk up.

"I don't want anything to happen. And if something did happen, I would go find them. That's the kind of person I am."

Keith Callis

Former U.S. Army tank gunner Keith Callis remembers heading into combat during the 1990-91 Desert Storm campaign to push Iraq out of Kuwait after the late Saddam Hussein's forces invaded the neighboring Middle East country and defied worldwide calls to withdraw.

"The ground war starts and we're rolling along blowing things up with the (M-1 Abrams) tanks, and these guys (Iraqis) are walking at us with their hands up. We just told them to keep walking," Callis, a 56-year-old Lampasas resident and native of Corpus Christi, Tx., said.

"We went in as far as Kuwait City. Remember the Highway of Death (six-lane Highway 80 between Kuwait and Iraq)? There are mines on the ground, and we're just straddling those so we won't get blown up. I see this guy on top of a BTR-50 (Russian armored personnel carrier), fueling it up, and I said to our tank commander, 'Hey, sir, there's a guy right there.'

"He said, 'Shoot the s.o.b.'

"So I fired a round and blew that vehicle up. Then we started going, and I'm looking around. I see this truck pulling out with a single-shot artillery cannon on it. This guy would back out, shoot one round, and pull back in. So, they could never find him. I saw him, and I told the tank commander. He said, 'Take him out.' So I did.

"My third shot, we were over at the Highway of Death trying to clear a ZSU-23-4 (four 23mm anti-aircraft machine guns) out of a bunker. It was really dug in, and we bombed and bombed, but couldn't get it. We sent infantry in, and the tank commander wanted to go up there, so we did.

"I looked over at the Highway of Death and there were 10, 12 (enemy) tanks moving down the road. I took a shot. It was 2,970 meters – roughly two miles. They were on the other side of the highway, and all I could see was turret. I'm tracking and I fire. It looked like the round hit the highway and ricocheted straight up in the air.

"I yelled at my re-loader to load another round, and we fired again. Right then, that turret started shooting flames and then it just blew off."

That was confirmed kill no. 3 for Callis, who served during Desert Storm from January to April 1991, the only combat deployment of his 12-year Army career.

Born and raised along the Gulf of Mexico in Corpus Christi, Callis grew up in a family of 13 kids headed by his U.S. Navy veteran father. He remembers some lean times growing up.

"It was hand-me-downs after hand-me-downs," he said. "Let's put it this way. When the youngest was born in 1970, we were living in a 900-square-foot house – with thirteen kids.

"Then they moved into an 1,800-square-foot house that belonged to my mother's sister and her husband. The girls had their own room; the big boys had their own room; and we had a garage converted into a bedroom."

After he graduated from Richard King High School in 1982, Callis was hoping to follow in two of his older brothers' footsteps and join the city fire department. That plan never worked out, and after working for a glass company and a stint in the oilfield, life served up a big, breaking curveball.

"I met a girl in 1987 and we got married two years later," Callis said. "She had two children, and the youngest was climbing a tree one day, and he fell. He was in severe abdominal pain, so we took him to the emergency room. They did an ultrasound, x-rays, some other tests, and they saw a bunch of stones in the pancreas.

"The doctor asked me what I did, and I said, 'Well, I work for the glass company from 7:30 in the morning until five, then I go to work at Builders Square from six to 11:30.'

"He told me, 'You know, you ought to join the military.'

"I said, 'Why's that?'

"He said, 'Your son's going to require a lot of medical attention, and it's going to cost you an arm and a leg.'

"So, I went in. I was already a certified EMT, because I had been trying to get on with the Corpus Christi Fire Department. I kept getting turned down, because I already had two brothers with the fire department and they didn't want a third. My dad was Navy; my brother, Pat, was Navy; my brother, Jim, was Navy. So what did I do? I joined the Army. Dad was pissed."

Callis enlisted at age 25 and reported for boot camp and advanced individual training (AIT) at Fort Knox, Ky., home of the infamous Heartbreak, Misery, and Agony obstacles.

"Ask anybody about those hills that you run," Callis said. "You look at them, and it's like, 'Oh, my God, we're going up that?'"

Callis not only survived boot camp and AIT, he excelled. By the time he reported to his first duty station at Fort Carson, Colo., in August 1989, he had been chosen to participate in the Excellence in Armor advanced training program and promoted from E-1 (private) to E-3 (private first-class).

During his first-ever gunnery practice at Fort Carson, Callis recorded the highest score in his company, and fourth highest in his battalion. That earned him a promotion to E-4 (specialist), just in time for the start of Operation Desert Storm (1990-91).

Following a rotation at NTC (National Training Center at Fort Irwin, Calif.), Callis and his unit was assigned to head down to Fort Hood to train National Guard troops preparing to deploy overseas.

"We were gone for 31 days, back for two days, then back out again for two to three weeks," Callis said, explaining that when he got back from Hood, he was slated for CQ (charge of quarters) duty at the barracks.

Then, a more interesting opportunity arose.

"I was at home and all of a sudden, my son is yelling at me upstairs, 'Dad, your first sergeant is at the door!'

"I went down there, and sure enough, there he was. I said, 'What's up, first sergeant?'

"He said, 'We need five guys from each company to send over to Saudi Arabia to be replacements, so when the ground war kicks off and somebody gets killed, their replacement will be there already.'

"I said, 'I'll go.'

"He said, 'Are you sure?'

"I said, 'First sergeant, I grew up watching war movies. I joined the service to serve my country. If I have the chance to go into combat, I'm going.'

"He left and my wife was standing there at the top of the stairs just looking at me, with the worst look on her face."

After Desert Storm, Callis was unexpectedly forced out of the military after suffering a severe ankle injury jumping down from a tank in the motor pool, being sidelined for several months as a result, and gaining 30 pounds. His ankle finally healed and he dropped the weight, but narrowly missed meeting the Army weight standard in time for his upcoming re-enlistment deadline.

After 12 years, in June 2001, the promotable E-6 (staff sergeant) was a civilian again.

"I loved the military. If I could have re-enlisted, I would have stayed and done my 20 or 30 (years)," Callis said. "I wish I had 20 years' service. There aren't too many jobs out there where you can work 20 years and retire with a pension – unless you're fire department, EMS, police department, a teacher …

"When 9/11 happened (Sept. 11, 2001 terrorist attacks), I tried to re-enlist. I went to the recruiter's office and said I wanted to go back in. He said, "Yeah, you can go back in, but you have to go to basic training, and you'll have to go back as an E-4.'

"I looked at him and I said, 'I did twelve years – I'm not going back to basic training.'

"That is my only regret – that I got out on the overweight program. It broke my heart. If I had gotten out under just a regular ETS, I would have been able to keep my rank and re-enlist. I was only out less than two months."

Nevertheless, Callis looks back fondly on his military career, and excitedly ahead to the future. After that initial disappointment, his dad came around, too, and was proud of not only his son's decision to serve his country, but also his later decision to volunteer for combat.

Callis said he and a brother bought 47 acres outside of Lampasas, and he plans to move out there sometime this year. He also stays busy with his duties as junior vice commander for the Lampasas VFW.

"I'm divorced now, since 2000. I told somebody, if I ever get married again, it's going to be on April 18. I joined the Army on April 18; I flew back from Desert Storm on April 18; my divorce was finalized on April 18.

"When I ETSed (left the Army) and I'm standing in front of my unit and they're giving me an award, saying thanks for your service, I told the guys, 'Some of you, I'll remember (and) some I won't, but I'm really going to miss that tank right there. Those things are a blast. Enjoy yourself when you're on one.'

"There ain't nothing like an M-1 Abrams. It was a lot of work, but it was a lot of fun, too."

Jennifer Meadows

Joining the U.S. Army was a childhood dream come true for Killeen resident Jennifer Meadows, but her once-promising career was cut short when a growing problem with alcohol finally caught up to the Michigan City, Ind., native.

"I got caught drinking in Kuwait," Meadows said. "I don't know about now, but back then, you could not drink at all."

The now 39-year-old teacher's aide was working as an Army food service specialist when her unit was sent to Kuwait in 2011. She worked long, hard hours but enjoyed every minute of it, even winning awards for her cooking. Beneath the surface, though, was a longstanding battle with depression and anxiety, and instead of taking prescribed medication, Meadows had a history of self-medicating with alcohol.

One day in Kuwait, she and some other soldiers went to a private party where the booze was flowing.

"It was like a fraternity party, and soldiers weren't even supposed to be there," Meadows said. "I love to dance, but in order for me to dance, I needed to have liquor in my system – liquid courage.

"So I was drinking and dancing, and it was time to leave. I went outside and was about to get into a vehicle, but there was another vehicle right behind me. I walked up to them – full of courage juice – and I asked them to move back a little bit, so we could get out.

"Turned out, they were the (military) police, and they smelled liquor on me. They were like, 'Come here.' I tried to hurry up and walk away, but they grabbed me and that was that."

The ending of a little girl's dream was underway.

Meadows joined the Army in 2007 while she was living in Germany with her military husband (now ex-husband). It was something she had wanted to do since elementary school back home in Indiana.

"I grew up with five older brothers and sisters," she said. "One served in the Marine Corps and two served in the Army. Me being the youngest, I always looked up to my older siblings, and they would come home on holidays, sitting in the living room, telling stories and bragging about their branch of the service.

"So even as a little girl, I always knew I wanted to join the military when I got old enough. After I graduated high school (in 2000), I was going to enlist, but I had gone to visit my sister who was stationed in Georgia, and while I was there, I met a soldier in a club.

"We wound up getting married, and he asked me not to go into the Army, so I didn't."

Eventually, that long-denied dream started coming back to haunt Meadows, and while the couple was stationed in Germany, she decided to go ahead and enlist. Her husband agreed and she was soon off to basic training at Fort Jackson, S.C., then AIT (advanced individual training) at Fort Lee, Va.

"I loved it," she said, of boot camp. "I loved the intensity of it. There were people who were crying, but I was not one of them. I grew up with a hollerin' mama, so it was no big deal.

"I was always athletic all my life, from elementary (school) on up – basketball, volleyball, track, all the sports. I always loved to be challenged, and my mindset was I wanted to become stronger. The whole time I was thinking I wanted to make a better life for me and my family."

While she was at Fort Lee, Meadows learned that her 44-year-old brother was dying from lung cancer back home. That was a blow, but she finished her training and was soon headed back to Germany.

She was well on her way to a promising new career, but her demons were also catching up with her.

"Now, I'm doing a lot of drinking. More than usual, since my brother just passed away. At the time, I was also still dealing with depression, but not wanting to get on the pills (anti-depressants) because you can't drink and be on the medication.

"Fast forward … I was getting a lot of awards for cooking. I had a great duty station. There was a lot of camaraderie, but I had been in Germany for a long time by that point. I wanted to be closer to my family, so I re-enlisted for Fort Hood because I heard it was a high-risk deployment station – where you get deployed to Iraq and

Afghanistan, because that's where I wanted to go. I wanted to fight for my country."

Both Meadows and her ex-husband came back to Fort Hood in 2010. He wound up getting injured at work and eventually was medically discharged. Meadows wound up in Kuwait and getting busted for drinking.

"I told them that I was thinking about killing myself," she said. "They thought I was just saying that because I was in trouble, but that was not the case.

"I was supposed to be on anti-depressants for many years, even before I joined the Army. But I chose alcohol to be my medicine. That wasn't my first time (being in trouble), but it was the last straw. They sent me to a rehab center first (two months in 2012), and then they put me out.

"I thank God now because, if I hadn't been in the Army, I wouldn't have been able to afford to go to a rehabilitation center. I would not have been able to afford to get help for my depression and anxiety. It saved my life."

Along with going to rehab in San Antonio, Meadows was doing lots of soul-searching, trying her best to find peace of mind and slay the dragons that had plagued her for years.

"I was watching a preacher on-line, and I used the things I heard him talk about. I made up my mind when I was in rehab that after I get out, I'm going straight to a church. And that's what I did."

When she made it back to Killeen, this time as a civilian, the first church she attended and ever became a member of was Destiny World Outreach Center. Something happened one Sunday that tested her newly growing beliefs, and also showed Meadows that she was finally on the right track.

"I was scared," she said. "My husband was getting out, so there was no money coming in. I was the breadwinner.

"One day at church, I put five dollars in the offering plate. I had to go out the next day and I went to get some gas, but my debit card wasn't working. I called the bank and they said there was no mistake. I had gotten paid the prior pay period, and I didn't have any money in my account.

"So I was driving, and I was talking to God: 'OK, God, I'm going to trust you.' Now, I'm thinking about different ways to get some money. Go home and look under the couches; look in the blue jeans pockets; under the mattress.

"Finally, my husband at the time called and said, 'Guess what? My money came in.' The money he was getting as part of his discharge hit the bank account that same day.

"From that day forward, I learned to put my trust in God."

Meadows has been alcohol-free since 2012, and off all anti-depressant medication since 2015. She became a licensed and ordained minister and now serves at Encourager of Life Ministries in Harker Heights.

"Discovering my relationship with Christ was the key.

"I tried for a long time to quit drinking on my own. Then I asked for God's help, and the temptation started to fade away."

A talented bowler who once rolled a near-perfect 275 game, Meadows is also a gospel rapper, writing her own music and songs. She says being forced out of the military was devastating at the time, but she is grateful for the experience and believes joining the Army ultimately saved her life.

"If that's what it took to get me to where I am today, then, hey … it was definitely a difficult experience, but I trusted the process that God was putting me through.

"If I had to do it all over again to become the person I am today, I would."

David Doell

Longtime Killeen resident David Doell served two tours in Vietnam, dodging enemy gunfire as a combat helicopter pilot, but the most terrifying time he ever experienced in the cockpit was during a routine maintenance test flight at Fort Hood.

"This was sometime in the '70s," said Doell (dell), a Texas native who grew up in the Canyon Lake area, between San Antonio and Austin. "Our mechanics always wanted to go up and fly. They worked on the aircraft, but they never had an opportunity to fly them.

"Every opportunity I had, I would always take a mechanic (for a ride). We had a new guy, and he was probably 17 years old, and there was an aircraft that came out of maintenance, and I had to (test) fly it. He came out there, and I asked him if he wanted to go up. He said, sure, so he did the inspection and signed off on it, and then I did my test flight inspection. When I did, I found a problem with one of the controls that goes to the transmission. It was just slightly out of tolerance, but I came down pretty hard on the kid.

"I said, 'You know, when you sign off on the paperwork, it should be good to go. There could be a major catastrophe from something like that. You need to check a little closer.'

"So we got up and flew, and one of the checks I had to do was a topping check, to make sure the engine is putting out max power. We had to go up to, like, four- or five-thousand feet, and I did all the checks and then we started coming down in a circular pattern.

"There was one cloud there – not a big one, but a cloud – and in the Cobra, I'm sitting in back and he's sitting in front. There was a rearview mirror so we could have eye contact, and as I'm coming down, I take the cyclic (control lever) and shake it just a little bit. The whole airframe just shuddered like all heck, and I see those eyes in the mirror get all big.

"He pushed on the intercom and said, 'What's wrong? What's wrong?'

"I said, 'I don't know.' And I went ahead and entered the cloud. He started screaming and was trying to get his shoulder harness off. He was going to try and jump out of the aircraft.

"I'm watching him, and I'm thinking, 'How am I going to explain this?' It probably scared me as much as it did him. We came out of that cloud and I got him calmed down. There was nothing wrong; I was just messing with the kid.

"I never pulled a stunt like that again."

Doell graduated from high school in New Braunfels in 1960, went to work, got married, had a couple of kids. He and his wife separated, and he got a letter from the military draft board.

"Uncle Sam said, 'Well, if she don't want you, son, I'll take you.' We got back together, but I thought it would be best if I went ahead and went into the military. I didn't really have a plan, or any big goals at that time. I was too young and dumb to set any goals.

"They tried to make a medic out of me, and I realized that after two years of that, I'd be so far in debt – I was getting sixty-four dollars a month – so I put in for flight school. Luckily, I was accepted."

Doell was drafted in 1964 and reported for boot camp at Fort Polk, La. He attended advanced individual training as a medic, went back to Fort Polk, then "pulled a few strings" and got stationed at the military induction center in San Antonio, a short car ride from his home in New Braunfels.

It was there that he received his paperwork to attend flight school, first at Camp Walters (Mineral Wells, Tx.) and then Fort Rucker, Ala.

"They said they were going to make me a maintenance officer, so they sent me to Fort Eustis, Va., and from there they were building up a unit at Fort Riley, Kan. I went to Fort Riley and we built a unit up there and we went on over to Vietnam in '67."

He served the first time in Vietnam in 1967-68, and again in 1970-71. He flew UH-1 Hueys and AH-1 Cobras.

"My primary mission was actually as a test pilot, but I also flew missions when they needed someone to fly missions," Doell said. "When I didn't have a lot of aircraft to work on, I enjoyed going out on missions.

"I got shot up eight different times. Never shot down. Just rounds taken in the aircraft. The pucker factor sure goes up – I can tell you that. You don't know where the rounds are going, (and) whether it's serious and if you're going to make it back or not. But I always made it back.

"I enjoyed flying over there. To us, it was a job that needed to be done.

"When you're going in (to a hot combat zone), if you were by yourself, it would be a lot worse. But when you're going in with four, six, eight, ten – or more – aircraft, you don't feel as vulnerable.

"The most I ever flew the first tour on combat missions was one day we flew thirteen hours and forty minutes (of) logged flight time. That was during the (1968) TET counter-offensive. Just keep taking them (troops) out there from one location to another, back and forth.

"You didn't really have a lot of time to think about it – you just did it."

Doell, now 79 years old, retired from the military as a Chief Warrant Officer 4 in 1986, when he was 45 years old. A Master Army Aviator, he received a Bronze Star and 14 Air Medals, and served not only as a helicopter test pilot but also a helicopter test pilot evaluator.

After leaving the service, he tried his hand at real estate sales for a while, then decided to go back to doing what he loves – working with helicopters for the U.S. Army at Fort Hood.

"I was putting in more hours in real estate than I ever did in the military; my weekends are all tied up. I had friends working in aviation on Fort Hood, and they talked me into coming back to work. So I did that, and I worked for DynCorp in aircraft maintenance.

"I was planning to work until I was age 60, but when I was 58, I told my (second) wife one day, 'You know, if I woke up one morning and I was dead, I'd sure be pissed. I'm thinking about retiring.'

"She said, 'Do whatever you want," so I retired again in 2000."

Five years later, Doell's wife, Sue, was diagnosed with Alzheimer's and so began a long, painful struggle with the debilitating and incurable disease. Sue wound up requiring around-the-clock care and died two years ago after the couple had been married 46 years.

"I took care of her until 2011, but I just couldn't take care of her the way she needed, so I had to put her in a nursing home. She was there for another eight years. I visited nearly every day. I only missed seven days out of those eight years."

Now, the father of three, grandfather of three, and great-grandfather of five, spends his time kicking around the house, and sometimes taking his boat out to the lake for some fishing.

His own future is somewhat uncertain these days due to medical concerns, but he continues to enjoy life and is grateful for his 22 years in military service.

"A week after my wife passed – two weeks after – I was having some shoulder pains, so I went to the doctor. Two hours later, they said, 'You have Stage 4 lung cancer.'

"Actually, it was Stage 3, so I've been dealing with that, and I think it's pretty much behind me right now.

"I have a bass boat, and I try to get out to the lake once a week, although I've been shut down the last three or four months. My neighbor always goes with me and he has some health issues, and I feel guilty about going fishing, if I'm not able to take him along.

"I am truly blessed. I don't have any financial worries, and I know most people do right now. That has to be terrible. But since that's not one of my concerns, it takes a big load off my back.

"I've got a lady friend, and we see each other a couple times a week. She doesn't want to re-marry; I don't want to re-marry. She has her life; I've got mine; and we've got ours together. So, it's a perfect world."

Bobbie Abbott

Oklahoma native Bobbie Abbott served 22 years in the U.S. Army, including two combat tours in Vietnam, and the 79-year-old father of two sons says the world he sees today makes him not only sad, but also more than a little angry.

"It's very upsetting," Abbott said, recently. "I think it sucks – pardon the French.

"I wanna cry and get mad as hell. People don't know what the heck they've got; they expect too much for nothing. I've got nieces and nephews who think they have to have everything when they first get married. Brand new cars; brand new house. I've worked 40 years, 50 years, for what I've got.

"If something don't happen real quick like, I think we may be coming to the end of the world."

Born and raised in Alva, Okla., Abbott grew up with four sisters and two brothers. His dad was a truck driver, and the family moved around some, but wound up back in Alva by the time he went to junior high and high school. In 1958, he decided to quit school and join the military.

"I had flunked the 10th grade English and couldn't graduate with my class, so I joined the Army.

"A guy I joined with had been in the National Guard and told me a lot about it. My dad was a World War II vet. They (his parents) thought I was crazy, but they went ahead and signed the papers."

In December that year, the 17-year-old reported to Fort Chaffee, Ark., for boot camp. A few years later, the Vietnam war began heating up, and he was sent in 1963 for a one-year stint as a fixed-wing mechanic and crew chief.

These days, memories of that historic time are a little hazy, but things evidently got off to a quick start when he landed in country.

"Well, I forgot a whole lot of it … but we went over as a platoon out of Fort Ord, Calif. We stopped in Hawaii and went to Guam, and then Okinawa and spent the night there, then landed in Nha Trang at dark, at sunset.

"Our aircraft got in – shipped over on the boat – and supposedly I got one of the first ones off the boat, after it was reassembled, and was a crew chief on it (a fixed-wing Otter). A friend of mine told me I was supposed to go out on a flight … there were two flights, (and) one was supposed to go out and pick up a dead captain, and the other one was a re-supply. The other guy says, 'I'm not taking the dead captain – you are.' I don't remember it, but I hauled the first dead captain out of the AO (area of operations).

"I flew the (Mekong) Delta, from Nah Trang to Saigon. In '70-71, I was a test pilot at Chu Lai. I left there right before the hurricane (Typhoon Hester) hit Chu Lai. My wife's father passed away and I was coming home on emergency leave."

Abbott retired from Fort Hood as a Chief Warrant Officer 3 and helicopter test pilot on Dec. 31, 1980. He went to school and earned two associate degrees, then joined the U.S. Postal Service, working 23 years as a rural letter carrier in the Kempner area.

Now, he spends his days noodling around on the computer, watching TV, and crafting wooden plaques and other artwork on his scroll saw. One of his pieces is to be displayed sometime soon inside the Copperas Cove Public Library.

"I make pictures and patterns. We (used to) have a bazaar out at Clear Creek Baptist Church on (FM) 2657, and I set up out there. They haven't had it in a couple of years, but they may start up again – I don't know."

As for the current state of the country and the world, Abbott says he fears very much for the future. He hopes things will get better soon, but sometimes he is not too optimistic.

"It could possibly be world war coming on. Too many of these groups want to take over everything. It's a scary fact," Abbott said. "I don't even want to go to Oklahoma and see my sister. I'm afraid of what I might run into.

"I think they need to start listening and stop protesting so darn much, and also to believe what they're being told. You're not going to get hurt, if you don't screw with the police when they stop you.

"I've been stopped several times here in Cove and the police have been the nicest they can be. I've had friends get stopped, and they haven't had any problems.

"I think maybe a lot of this all started back when the government started saying you can't spank your kids, or even discipline them in any way. And when there started being so many single mothers out there, with no fathers in the home.

"As far as the COVID goes, it really hasn't changed anything in my life. I'm a quiet, laid-back guy. We (he and his wife) don't do much. The main thing it has done is kept us from going out to dinner and things like that. We used to go out a lot, but now we can't just pick up and go somewhere without putting on a mask … which doesn't bother me, except it's awful hot.

"We were in Arizona (visiting his sister) when this thing hit, and we had to leave real fast like, because they were going to close down the state.

"I think a lot of it is a bunch of hogwash, trying to take over the lives of everybody. If you stop and think about it – is it really this bad? You just have to take it one day at a time, and hopefully you don't get it, and nobody that you know gets it, until it finally dies out."

Rebel Brown

Some people say they have worked "all their lives," but U.S. Air Force veteran and Gatesville, Tx., native Rebel Brown may fit that description better than most.

"The first job I ever had, I was in second grade," the 72-year-old former staff sergeant said. "During the summer, I went tent to tent at North Fort Hood selling newspapers. I made one penny per paper.

"I was so short that I drug the newspaper bag around, and those soldiers would see me coming and feel sorry for me. They'd buy a paper for a nickel and give me a dollar and not take the change. Sometimes it would be a nickel extra; sometimes maybe a quarter.

"There were times I'd sell fifty papers (and) sometimes I'd be known to sell a hundred. As a result, I would make quite a bit of money. I'd save all of it and buy my school clothes for the year."

Born Feb. 19, 1948, Brown graduated from Gatesville High in 1966, attended Texas A&M University for a year as a member of the famed Corps of Cadets and Fightin' Texas Aggie Band, then joined the Air Force. With four kids in the family and dad working as a mechanic and service station owner, going to college for four years was a little out of reach financially.

"There were three of us boys and one girl, and we couldn't all be in college at the same time," Brown said. "My older brother went to college for a year and then went into the Air Force. I went a year, and then my younger brother went two years, and then he went in."

Although it was short-lived, his college career helped him when it came time to report for boot camp in 1968 at Lackland Air Force Base in San Antonio.

"After being at A&M, boot camp was easy," Brown said. "Being in the Corps was rough. There was a lot of hazing back then."

Based on the results of aptitude tests, Brown specialized in electronics in the military, attending school at Keesler Air Force Base in Mississippi, then serving at Holloman Air Force Base in Alamogordo, N.M. He remembers being stuck at the airport in Albuquerque and hitching a ride to his new post with an officer also on his way there.

"I was going to fly down to Alamogordo, but I couldn't get a plane out. I was in uniform, and this major came up and said that if I was going to Alamogordo, he would give me a ride. He had the same problem, but he had an Air Force plane coming up for him. So, we got a ride that way.

"I was in ground-to-air communications there. They sometimes called us the 'junk men,' because anything that came up that had to do with electronics, that's what we worked on. I worked on everything from church organs to facsimile machines for the weather."

Part of Brown's service also included time working on classified systems, which he still is hesitant to talk about in any detail.

"Some of it I can talk about, and some I can't. Since I was in the electronics (part) of it, some of that stuff they still use, so I try and not talk about it too much."

As the U.S. involvement in Vietnam started to wind down in the early- to mid-1970s, Brown left the service after seven years and headed back to central Texas. He went to McLennan Community College in Waco, where he earned an associate degree, then to the University of Texas at Arlington to study chemistry, then back to Gatesville for a while. He worked at a grocery store, a television repair shop, mixed drinks at a couple clubs, and spent some time with Texas Instruments in Dallas.

One day, he saw an advertisement for substitute teachers in Copperas Cove, and after volunteering as a reading teacher for adult literacy programs in Temple, Brown – with a little nudge from his sister – decided to apply.

"My sister is a retired teacher, and she talked me into going over there and applying.

"I started working with (special needs) students, and once they found out I would work with the special ed(ucation) kids, I worked as much as I wanted to. I wouldn't even wait for an assignment every day. I would just go up there and see which classroom they wanted me to go to. Finally, they just put me on full-time."

After 14 years with the Copperas Cove school district, Brown retired two years ago at age 70, and now spends a lot of his time reading. He is a member of a number of different military organizations, including Disabled American Veterans, American Legion, Sons of Confederate Veterans, and Sons of the Republic of Texas.

Brown was recently chosen as a recipient of the United Daughters of the Confederacy National Defense Medal, which is awarded to "lineal descendants of Confederate military and who have given a minimum of 90 days honorable active-duty military service to our country."

His family moved to Coryell County from Tennessee in 1854, and his ancestors have been traced back as far as the Revolutionary War (1775-83). Some fought in the War of 1812 and Texas Revolution. His great-great-grandfather fought in the Civil War, something he does not advertise, but is proud of nonetheless.

He is named after one of his great-grandfather's sons, Rebel John Brown, who once served as a doctor on Fort Hood.

"His house burned down in 1922 – Oct. 9, 1922. There was like nine kids, and daddy lost two sisters in that house fire.

"They moved to Gatesville, and everybody knew him as Dr. Reb. When I was going through school, everybody called me Reb. When I graduated, they read my name as Rebel. That's the first time some of my classmates heard it.

"When I went to A&M, and to the service, everybody had to use their full name, and that's when I started going by Rebel. My oldest brother is Robert Lee Brown.

"All this controversy now about the confederacy … my great-great-grandfather was in the Civil War. Everyone in Coryell County fought in the Civil War. They didn't have slaves. They weren't fighting for slavery; they were fighting for their land.

"They talk about (confederate) statues. I don't care who the statue is of, or what it's for, you should not destroy it. It's somebody else's property. Some of the things that's going on should not be allowed. You don't go painting buildings. There's laws against it. If I go down here to the courthouse and paint the streets in front of the courthouse, I get arrested.

"There's legal ways to take care of things like that, if you don't like it."

Michael Weisheit

Michael Weisheit served two tours in Vietnam and was wounded three times – shot in the arm during an ambush, hit by some shrapnel, then blown up and nearly killed by an enemy grenade as he and two other soldiers drove what back then served as a highway mine sweeper.

"We called them Polish mine sweepers," Weisheit said. "What it was, we used five-ton Army dump trucks, and we pushed the window down, took the headache board off, filled the truck with sandbags and dirt, and backed down the highway. Just put it in reverse and backed down the highway.

"We didn't have any of these fancy mine sweepers like they have now. That's how we did it. It was effective, and we hit a lot of mines, but never had anybody killed from it. Most of the time, it blew the back wheels off the truck, but nobody died.

"Toward the end of my (second) tour – I only had a few weeks left – we moved into Da Nang, and we were doing some mine sweeping, and we got ambushed again. This time, they threw a hand grenade in my truck before I could get out.

"It killed my friend, who had three weeks left. It landed right behind me, between the bed of the truck and the cab. If it had landed in the cab, it would have killed me. It blew upward and caught me in the top of the back. Put two holes in my left lung and blew a couple of ribs out. Blew part of my shoulder blade off.

"I knew what was going on, but I couldn't breathe. I knew if they'd get me on that chopper, I'd be OK. When I saw the smoke blow, I knew the chopper was there, and I knew I'd be all right.

"When I got to the hospital, the doctor cut my throat to give me a tracheotomy, and that's the last I remember, when he did that."

Weisheit, who lives now in Killeen, was born in San Francisco, moved to Davenport, Iowa, when his U.S. Army father – who served with the Coast Guard during World War II – went to fight in the Korean War (1950-53). His dad died in combat when he was four years old, and he wound up three years later in Tucson, Ariz., when his mother re-married.

He graduated high school in Tucson, and was drafted by the military in 1968, when he was 19.

"I was thinking about going in anyway, but I got my draft notice, and so I went ahead and volunteered," Weisheit said. "I couldn't get a job because, in the '60s, if you weren't in college, you couldn't get a job. Nobody was going to spend any money on you because you were going to get drafted. Everybody was getting drafted back then."

Six months later, he was a combat engineer clearing minefields and building firebases in Vietnam. About six months after he arrived, an ambush during a mine-sweeping operation left him badly wounded, with a hole in one arm and severed tendons to all his fingers. Five surgeries and six months later, he was sent to Fort Carson, Colo., and then back to Vietnam.

"I didn't mind going back," Weisheit said. "I was pretty stout back then. I was young and tough, and I thought I knew everything.

"I went to A Company with the 26th Engineers at a place called Hawk Hill, which was north of Chu Lai, right near Highway One. Our main job was building the fire bases. You build a fire base out on one of these hilltops in the jungle. You blow the top of it off, and then the engineers clear it and get it set up, and then you bring howitzers in – 155s, 105s, and mortars.

"Then the grunts come in – the infantry – and they go out on patrol, to find the enemy and kill 'em. They can only go out as far as we can support them (with artillery), which was about 17, 18 miles. When they've cleared that area, we'll go build another fire base about 25, 30 miles away."

Weisheit wound up with three Purple Hearts from his time in Vietnam. He was medically discharged after recovering from the grenade blast and went on to work 17 years for the federal prison system, and six years for the U.S. Postal Service.

With his 72nd birthday coming up in a few days, the father of nine and grandfather of a dozen or so is retired and 100-percent disabled. He comes from a long line of family members serving in the armed forces, including his grandfather, father, uncle, father-in-law, stepfather, son, and two granddaughters currently in the Army.

He is proud of that heritage, and proud of his time in the service.

"I was planning on staying in for a 20-year career, but they discharged me," Weisheit said. "I enjoyed it. I never met with any of the protests or anything back then, because I was medevaced both times back. I came back on medevac aircraft.

"When you're in combat, it's maybe for five, 10, 20 minutes – most of the time. That's it. The ground shakes … when you've got two platoons (and) 50 guys with M14s (later M16s) all firing their weapons.

"Then, it's boring for a month. It's not like they show on TV, where it's constant fighting. That's not it.

"When I got hit in the arm, we were in a little valley, and we thought they were all out in front of us, but they were behind us, too. That's how they got me. I saw the bullets coming. They were hitting the ground around me, and I jerked to get out of the way, but they caught me in the arm.

"They tried to finish me off, but I mentally made myself small. In my brain. That way, they couldn't hit me. Probably 25 bullets came close, but none of them got me. That was totally scary.

"I was never political. Back then, when Uncle Sam called, it was your duty, and you felt honored to be able to go."

Will King

Growing up in small-town central Georgia, Will King always knew he wanted to be a soldier.

After graduating from high school, he took a little time to relax – "I hung out and partied and drank for about six months." – then pursued his lifelong dream to join the Army.

"It was just something that was always in me," the 50-year-old Killeen resident said recently. "The military mindset just fell right in line with my personality and the way I viewed the world.

"Even as a child, I was the same way. I just knew there was a right way to do things … that things needed to be done in order, and in a procedural way. It was mostly out of being a protector. I'm a natural protector.

"I always knew I was going to join the military. I always tell everybody that I was a soldier way before I went in – I was just waiting for time to catch up."

Born and raised in Warner Robins, Ga., in a family that included nine children, King reported to Fort Benning, Ga., in January 1989 for boot camp and advanced infantry training, then went to airborne school. His first duty station was Fort Richardson, Ak., where he served four years, then headed down to Fort Campbell, Ky., where he worked at the air assault school.

"That was an interesting time," King said. "I was still partying a lot, and I got in a lot of trouble at Fort Campbell. I was a bit wild and mouthy back in the day."

After Campbell, it was back to Alaska, where he participated in the biathlon (skiing and shooting) in the Winter Military World Games, then on to Fort Bragg, N.C., and the famed 82nd Airborne Division, where King injured a knee and was re-classified to radar technician, after 13 years in airborne infantry.

That was the role he played during his later deployments to Afghanistan (2003-04) and Iraq (2005-06).

"Anything with electricity running through it for us and our coalition forces – anything that broke, they brought it to me," King said.

"Afghanistan wasn't that bad. We got the rockets (attacks) and the usual stuff, but it wasn't as bad as other places. I might be putting it a little bit lightly, because I'm combat-minded, but it wasn't nearly as bad as it could have been.

"You had the constant threat of death, just like everywhere else, but I didn't think it was as bad as it could have been. There were a couple of times when they got kind of close, but it was still pretty mild, in my mind, anyway.

"They attacked us and blew some stuff up, set some stuff on fire. Had to get in the bunker a few times … but I didn't lose any of my guys in Iraq or Afghanistan."

Retiring as a staff sergeant in 2011 following 23 years' service, King also saw duty at Redstone Arsenal in Alabama; Fort Sill, Okla.; Fort Drum, N.Y.; and Fort Hood.

Now working as a programs coordinator for Hill Country Community Action Association, the father of two daughters who is also an artist, musician, owner of "Fit for a King Bakeries," and heavily involved in his church says he enjoyed everything about his military career – almost.

"I enjoyed the training; I enjoyed the discipline; the structure; the physicality of it; the challenge; the whole thing. Jumping out of airplanes is not something most normal people do. I trained with probably every special forces branch of almost every country we are allied with. That was fun."

One of the most memorable experiences for a lot of paratroopers is that first parachute jump. King is no exception. He says he remembers that day vividly.

Well, most of it.

"The first time I was on an airplane, I was jumping out of it, and I had the door (first to jump)," he said. "When they said, 'Go,' I can't tell you what happened for about three seconds after that. I blacked out for about three seconds. After that, it was all good. I'm still here, so I guess it worked out."

King decided to leave the Army after several years of what he saw as a decline in morale and leadership quality. He has some advice for young men and women considering an enlistment, or those who are just starting out.

"Know what you're going in for and believe that. The military is not a job. It is a lifestyle. It is not somewhere that you just want to go to pay some bills or go to college. If you're not 100 percent in, do not go – because you might get somebody killed.

"If you are not fully aware of what the military's purpose really is, and your place in that, then do some homework. Do some research; do some soul-searching. It's about whether you have the intestinal fortitude to do what is going to be demanded of you, regardless of which job you get assigned.

"I always told my soldiers, 'Your M.O.S. (military occupation specialty) is what you do in between fighting wars.' At any moment, your primary purpose is to protect and defend. That means you might have to pick up a rifle and fight. You are a fighting force. All this other stuff just keeps you busy until it's time for us to go to war.

"It's not just a job. It takes over everything that you are, and if you're not 100-percent bought into that, then you're going to have a problematic career."

Edward Smith

Former Fort Hood combat pilot Edward Smith was minding his own business going to college in west Texas when a letter from Uncle Sam put an end to his studies and changed the course of his life.

"I was going to college because that's what everybody did then," the 82-year-old from Hale Center, Texas, said, recently.

"I really didn't have anything in mind. I had no idea what I was going to do, until I got drafted (in 1961). Then, after I got into the military, I decided, 'Hey, this is pretty nice. I think I'll just stick around for a while."

Smith stuck around for 20 years, retiring in 1981 as a major and serving two tours as a reconnaissance pilot in Vietnam.

He went to boot camp at Fort Carson, Colo., stayed there for two years, and then with his background in college and ROTC, applied to become an officer and was accepted for a direct commission. He later served at Fort Benning, Ga., and as a company commander at Fort Leonard Wood, Mo., then applied for flight school.

Soon, Vietnam came calling.

"About a week before we graduated, they sent us all to Vietnam," Smith said. "I didn't worry about it. It was all so new; it was just another assignment."

He arrived in country the first time in June 1965 and stayed for 18 months, logging 2,000 hours flying recon missions in the famed low-altitude Bird Dog aircraft. There were a number of close calls searching the jungle for enemy troop locations, but Smith said he never felt afraid.

"I was concerned several times during both tours, but I was never afraid. I always thought I was going to come back, and I did.

"I was over into Cambodia one time, and we found a big encampment … they didn't like that at all. I came back with quite a few bullet holes in my plane. But that ol' plane brought me back. Different times; different things happened. I never worried about it. Whatever they wanted us to do, away we went. There were times I probably shouldn't have, but I did.

"When we first got over there, it was all so new, and the buildup was going on. They told us to fly over at a thousand feet. We saw right away that at a thousand feet, you can't see anything … in those jungle and forest areas. So against orders, we dropped down to treetop level – and then we could see something. Of course, they (enemy) could see us, too.

"We started working two-plane recons. One would be at a thousand feet, and the other would be right on the deck. The one at a thousand feet would keep up with you, and if you saw anything, you told him, and he'd mark it on the map. Then, we'd either put artillery in there or call in an air strike. Whatever we could get."

After another tour in 1970, this time for only seven months, Smith came home and was assigned to four years' duty in Germany, followed by a stint at Fort Benjamin Harrison (Indiana), and then Fort Hood, where he retired and went to work as a training coordinator in civil service.

He held on to that career for decades, finally retiring for good at age "72 or 73." The father of three and grandfather of four then hit the highway with his wife of 53 years, Beth, to do some traveling, but arthritis in his back has kept him off the road for a while now.

He looks back with pride and satisfaction at his military service, Smith says, including Vietnam. Like a lot of other vets who served in that controversial 20-year-long conflict that cost 58,220 American lives, Smith says he was resentful for a long time at the lack of respect and gratitude he and his buddies received.

"When Vietnam was going on, we went over as individuals, and came back as individuals. When we came back, the Army was dropping its numbers quite a bit, and there was a lot of guys who had 12, 15 years in (and) two or three tours in Vietnam, and when they came back to the States, as soon as they got here, they (the Army) said, 'OK, you're released.'

"They put 'em out on the street. That was pretty hard to take. They never gave me that opportunity," he said, laughing, "so I just stayed in.

"When we came back, there was no parades. Nobody met us and slapped us on the back or anything – except my wife and my parents.

But we never thought anything about it, until when everybody came back from Desert Storm, they had all the big whoop-de-do there.

"We said, 'Wait a minute. They went over there for a hundred days; we went over for ten years, and we got no recognition.' That's when they started saying, 'OK, maybe we ought to start saying something about these old boys.'"

Recently, Smith was one of six area military vets to receive Quilts of Valor from TX Patriotic Peacemakers, in honor of their service.

"My whole family was there – my three kids; my wife; my grandkids; and a bunch of my friends. It made me feel pretty good. I wasn't a big hero. I was just another old soldier out there doing his job the best he could.

"In Vietnam, there were people coming over there who kept up with every day they did anything, and wrote themselves up for all kinds of citations and medals … I never did do that. I just went out there and did my job. If somebody wanted to put me in for a medal, well, ok, that was all right."

Along with his Bronze Star, Air Medal, Meritorious Service Medal, and several "I Been There" (Vietnam Service) medals, Smith recently received another unofficial honor that in some ways outshines all the rest.

"A friend of mine who we went to flight school together and to Vietnam together … he spent a year over there and then came back and got out of the Army and went on with his life.

"He got a Distinguished Flying Cross, and he came by here the other day, and he said, 'You know, you deserve this a helluva lot more than I do. I'm going to give it to you.'

"I told him, "You giving me this medal means a lot more than if some stiff-neck general did.' And it did. It made me feel real good."

Larry Shatto

Retired Apache helicopter pilot Larry Shatto was one of nearly 300,000 young men drafted for military service in 1968.

With the Vietnam war raging (16,592 American casualties were reported that year; the highest annual number), that order all but guaranteed a trip into combat, but young Shatto never looked back as he headed straight away to the recruiter's office.

"It was a year after my eighteenth birthday," the Vallejo, Calif., native said. "I was invited by the U.S. government to take my draft induction physical.

"I knew enough that taking that physical meant I was either going to be in the Marines or in the Army, so I figured I'd go into the Army. I ran down and enlisted within a day or two with a buddy of mine from a nearby town.

"There were guys back then trying different tricks to get out of it – you could play the gay card, which worked back in that day. Or, for instance, I had a buddy who had psoriasis (a chronic skin disease), and what you'd do is, a week or two out, you did everything you could to aggravate that condition so that when you go in, you're declared 4F (unfit for service).

"Another trick I heard about was, if you eat a whole lot of certain foods, like bananas, it drives your potassium level up, and you can fail the physical. There were all kinds of little tricks to fail a physical.

"There were five of us buddies who all hung together, and out of those five buddies, one got a married-with-two-kids deferment; another one was going to college and got a school deferment; a third was 4F because of something wrong with his feet – flat feet or something – and a fourth buddy went into the Navy. I was the only one who went into the Army."

Now 71, Shatto spent 11 years as a helicopter mechanic and crew chief, becoming a staff sergeant before earning his Army aviator wings as a warrant officer in 1980. He served in Vietnam from January 1969 to January 1970 as a crew chief in the 176th Assault Helicopter Company in the Americal Division.

Shatto says he was shot down twice during combat missions aboard gunships and smoke ships, including one frightening crash seven months after he arrived that earned him a Purple Heart.

"On 13 September 1969, we were covering a medevac that was going into an observation post, sitting on top of a ridgeline. We had been fighting on that particular hillside for, like, 18 hours. We were in the gunships, so we had made four or five trips out there – shoot 'em up, bang, bang – all night long, and this was the next afternoon.

"We were going in there, and this guy stepped out from behind a tree – we called them 'one-shot Charlies'; the guys who would jump out of a tree line or a bush, fire one shot and then take off and run – took one shot and got lucky and took our engine out.

"So, we auto-rotated, and … you've seen a dodo bird? It looks all beautiful and graceful coming in, then as he lands, he's going too fast, and he puts his feet down and from that point on, it's just a terrible tumbling, rolling, flipping, bouncing crash. Well, that's what we did in our helicopter.

"We bounced about four times, 10 or 15 feet in the air, rolled over, and the pilot busted his lip and bit his tongue in half; the co-pilot cut the livin' daylights out of his arm getting out of the aircraft. We had a mini-gun – basically a six-barrel Gatling gun – that sits on the left side about three feet from the crew chief's legs, and then a seven-shot rocket pod just below that. When we bounced and landed, those systems busted off and the mini-gun hit me in the leg.

"All it takes for (the mini-gun) to go off is for the barrels to rotate, and had it gone off, it would have blown my leg off and killed the pilot. It cut my leg a little bit. The guy that got really hurt was the door gunner, who got thrown out of the aircraft. He had his seat belt on real loose, and he had a bruise all the way from his shoulder blades down to his butt. Almost broke his back, but the aircraft rolled over on his leg and his foot.

"Two of us lifted up on the aircraft enough to get the weight off his leg, and then we drug him out to a waiting aircraft that had landed a minute or two before us. The three of us were back flying within two or three days, and then Hank was medevaced out to Japan."

After post-war stints at Fort Sill, Okla. (where he met and married his wife of 48 years), in Panama, Davison Army Airfield at Fort Belvoir, Va., and then graduating from flight school, Shatto went to Fort Ord, Calif., where he served as a helicopter test pilot. He continued testing AH-1 Cobra helicopters in Germany from 1982-85, and then was reassigned to Fort Hood, where he was a test pilot for the 6th Air Cavalry Brigade.

After retiring from the Army in 1989 as a Chief Warrant Officer 3, he went to work as a civilian contractor, and participated in Operation Desert Storm in 1991, doing maintenance test flights on Cobras and Apaches in Saudi Arabia, and then back home to Fort Hood, where he quit flying in 2015.

These days, the father of two, grandfather of seven, and great-grandfather of two continues to support the troops as a dedicated volunteer for such organizations as Snowball Express, which serves children of fallen military heroes, the American Soldier Network, and the Tribute to Valor Foundation.

He is particularly close to America's Gold Star Families, an organization that provides "honor, hope and healing to those grieving any military loss through active-duty service in the United States Armed Forces."

"I fully support, defend, and will do anything under the sun with regards to Gold Star Families," Shatto, a Harker Heights resident, said.

"I refuse to stand by and allow the American public to treat the current generation of warriors like they treated the Vietnam veterans. These soldiers and their families are America's best, and they deserve our upmost respect.

"You can say anything you want about the military – we're killers; we're mercenaries; we're terrible people – but do not disgrace me and my service when I'm standing next to the mother and two kids of a soldier who died for your right to say that."

Max Constant

Max Constant was in the eighth grade in Hutchinson, Kan., when Islamic terrorists crashed commercial airliners into the World Trade Center towers in New York City on Sept. 11, 2001.

At first, he did not understand what was happening. In fact, he thought for a while that what he was seeing on television must be some sort of action movie with excellent special effects. After he learned what really occurred, the teenager decided then and there that as soon as he could, he was going to join the military.

"From the time I was a kid, I always wanted to be in the military, wanting to get out of a small town, and 9/11 was kind of the major push," Constant said. "I was only in eighth grade, but it kind of solidified everything for me.

"Our middle school started at 9 o'clock. I think I was getting ready (for school), and I saw it on the TV, but didn't pay too much attention to it.

"When I got to school, I was talking to some buddies. This was right when the Spiderman movie was supposed to come out, and one of the clips showed Spiderman with this big web stuck up between the Twin Towers, and he caught a helicopter that way. So, I was thinking that was some pretty good graphics they had, because the buildings were falling down. I thought it was kinda cool and laughed a little bit, not knowing that actual people were dying.

"Miss Robinson was the teacher, and she started chewing me up one side and down the other, and I was like, 'Hey, it's a movie. Nobody dies in movies.'

"She turned on the news and that just kind of … I had so many emotions going through me, and I knew I had to do something to make up for that mistake I made."

Constant enlisted on June 1, 2006, when he was 19 years old and reported to Fort Leonard Wood, Mo., for boot camp. Now in his 14th year of active-duty, the 33-year-old staff sergeant combat engineer is stationed at Fort McCoy, Wisc., with his wife of nine years, Christina, and their two boys, Kyle, 5, and Kole, 3.

He served at Fort Hood with the 36th Engineer Brigade from December 2006 to August 2011, transferred to Fort Bragg, N.C., and then returned to Fort Hood from November 2013 to October 2018, serving with the 1st Cavalry Division.

His overseas assignments include two nine-month rotations to Korea, and a pair of deployments to Afghanistan (2008-09; 2012-13). His duties in the Middle East included road construction and convoy security during LOGPAC (logistics package) missions ferrying supplies to various bases.

Aside from being "hot in the summer and cold in the winter," Constant remembers a strong camaraderie in his unit that helped him and his buddies handle a number of close calls during some dangerous situations.

"Lot of firefights," he said. "Them popping shots on us; dropping mortars; stuff like that. We had some great air cover boys with us, so they tended to kind of scare everybody off.

"At the time, it's just an adrenaline rush, and you do what you've got to do to get back (safely). You've got to protect yourself and protect your buddies to your left and right. Later on, that's when you kind of replay everything, and then you realize the 'oh, s—t' moments.

"I was fortunate enough to be in a very tight-knit platoon, and you could kind of see it come over dudes' faces, and you give them a minute and then you give 'em that good hard nudge or punch in the shoulder, or something like that, and kind of joke about it and bring them out of that funk. It's just a bond that you build.

"I have a lot of great friends I've served with, and we all get scattered to the four winds once our time is up at certain bases, but the bond you have with guys you've been through things with is unbreakable."

Constant says he plans to serve six more years, then come back to central Texas (his wife's family lives in the Killeen-Harker Heights area) and pursue a career in law enforcement. He enjoys the Army and has no regrets making that decision way back in eighth grade.

"You have your good days and bad days, but if I had the chance to do things over, I'd do it again," he said. "If I could go back, knowing what I know now, I might have done a couple things different. But at the end of the day, I'd still be in the Army – 100 percent.

For young men and women thinking about a career in the military, Constant has this advice:

"If you're considering it, just do it. You'll find out quickly if the military lifestyle is for you or not. But you have to give it a chance.

"Guys who are in, who are thinking about just doing four years and then quitting, to me that was never the answer. If you don't like where you're at, try to go somewhere else. Pick a different duty station and see what's there. There's a bunch of different worlds out there than the world you're in.

"Me, I'm an airborne guy. I went to airborne school right after AIT (advanced individual training); went to Fort Hood; went to Fort Bragg; and Fort Bragg changed my entire outlook. Fort Bragg was a blast. I loved it. I loved jumping out of airplanes. It was a lot of fun.

"As a young soldier, you'll see things that you don't really like, but that's why you should keep going; keep doing what you have to do in order to reach a position where you can change the things you don't like for the better."

Don Nicholas

Retired Army National Guard master sergeant/first sergeant Don Nicholas never saw combat; never served an overseas deployment during his 31 years in the military. But he excelled at a job that many soldiers could not or would not want to do.

"I see guys with Silver Stars and Purple Hearts, and I've met Medal of Honor winners," Nicholas said. "And when I say I was a recruiter for 27 years, they look at me and they say, 'Man, I wouldn't trade places with you for nothing.'

"They've told me, 'You wouldn't want to do what I've done, and I really don't want to do what you did.'"

Now 68 years old, Nicholas grew up milking cows, baling hay, slopping hogs, feeding chickens, and picking up eggs with an older brother and sister on the family farm in North Dakota before deciding to join the National Guard in 1971, as U.S. involvement in the Vietnam War was beginning to wind down.

He was going to college at the time, earning a degree in animal science and agricultural economics – even teaching one year of high school agriculture – along with his part-time Guard duties and never-ending farm chores.

"I figured I'd do my time, like a lot of folks did, and then get out and go back to doing what I was doing," he said.

"Then, I fell in love with it. I was looking for an opportunity to go on active-duty and in 1978, when I was 26, they came up with this AGR program – Active Guard Reserve – and it's still a great program today. A lot of Guard people are in it now.

"I was one of the first people hired in 1978 under Title 10. I was the same as every active-duty soldier, sailor, airman, and Marine at that time, for about eight years. Then, they switched us over to what they called Title 32, which puts us under control of the governor at that point."

Initially, Nicholas was a combat engineer before becoming a recruiter. He worked at the Military Entrance Processing Station (MEPS) in Fargo and was responsible for a large territory that included the state of North Dakota and northwestern Minnesota.

It was a demanding job with sometimes grueling hours, but also something Nicholas enjoyed. Not only that, being a National Guard recruiter eventually led him to a dream post-military retirement job.

"A lot of guys go into recruiting and it's not a good career for them. They have a hard time with it," he said.

"It becomes a downfall for them, and they look at it as a black mark on their record if they're not successful at it. It can make or break your military career, sometimes. It's long hours, and it's essentially a sales job.

"You have to maintain a certain level of composure all the time; positive attitude all the time. You have to work really crazy hours. It's not uncommon to get up at two or three o'clock in the morning to go pick up an applicant and take him overnight to a military entrance station – and those are not in every town.

"In North Dakota, we only had one military entrance station, so if an applicant lived in Bismarck and was going to enlist, he had to come in the day before, and the recruiter was responsible for driving him (to Fargo).

"The recruiter would have to pick that applicant up, make sure all the paperwork was done, get everything put together. You put in about eight to 10 hours' work just getting them ready. That doesn't count the number of hours you've put in trying to get the individual to sit down and visit with you first. Not everybody is a walk-in. Not everybody walks in the door and says, 'I want to join the National Guard.' You have to go search them out."

After he retired from active-duty Guard in 2003, Nicholas went to work as a recruiting contractor for Management Training Consultants Inc., in Washington, D.C. He and other recruiters traveled the country to find and sign personnel for the Ground-Based Midcourse Defense System and 49th Missile Defense Battalion at Fort Greely, Ak.

"That base was closed, and that was the reason they re-opened it," Nicholas said. "We were the ones who put the first people on the ground up there; first boots on the ground. Three of us traveled for almost four years all over the United States recruiting people who were in the Guard to go on a three-year active-duty tour to Fort Greely."

In 2006, MTC moved to Killeen, and he and his wife, Jody, who also worked for the company and now serves on the Harker Heights City Council, came to central Texas. Four years later, Nicholas went to work as district director for former State Rep. Ralph Sheffield of Temple, and in November 2014, he got a call from U.S. Rep. Roger Williams, R-Texas, asking if he was interested in a job on the congressman's staff.

He jumped at the chance and has worked for Williams ever since.

"That was kind of my dream from my high school days – to work as a field representative or staffer for a congressional representative," Nicholas said. "I never wanted to be an elected official; I just wanted to work for one."

Looking back at his military career and ahead at the future of the country, Nicholas says he is proud of his service, and excited about the possibilities to come.

"I'm extremely satisfied with my career," said the father of seven, grandfather of nine. "In some ways, I wish I would have stayed on and made a deployment to Iraq, but I never did. I was a master sergeant, a senior NCO, and I was ready to retire. I had over 30 years in, so it was time.

"I had been selected to go to sergeant major academy, but I decided I wasn't going to worry about going to another nine-month school, and so I took the opportunity to work on the missile defense program. I figured if I missed it, an opportunity like that would never come along again.

"I think recruiting was a career that gave me that opportunity … it helped me understand there's more to life than small stuff. Really, it's all small stuff. You look at it as all small stuff and you take the small steps to get to the level you want to be."

David McClure

Retired U.S. Army Lt. Col. David McClure came of age during the Vietnam War era, but the Maryville, Tenn., native had no intention of ever serving in the military.

Not only did he wind up wearing a uniform and leading infantry soldiers into jungle combat, McClure served a total of 30 years, including four years' active-duty and 26 years in the Army Reserve.

It was early 1968 when McClure graduated college and was preparing for a career in education, teaching science in his adopted hometown San Antonio, where he grew up and went to high school. Instead of setting up his new classroom and writing lesson plans, however, he found himself later that year lacing up a pair of combat boots.

"I couldn't afford college, so I went to work as a plumber for a while (after graduating high school in 1962)," McClure said. "I was a union plumber's apprentice. I worked on two different hospital jobs – one in Kerrville, and then Scott and White in Temple (1962-63). From there, I finally saved enough money to go to community college – San Antonio College – and that was when I had my first indoor job ever, working at the county tax office, as a deputy collector of taxes.

"I went up to Southwest Texas State College – which is now Texas State University – in San Marcos, where I finished my degree in January 1968. I had worked as a plumber for the college maintenance department – for one dollar an hour – and then a better-paying job as a residential advisor at the Gary Job Corps Center. The plan was I would continue working as a plumber until the fall, when I was going to start teaching and coaching in San Antonio.

"They put in for a (draft) deferment from the school district, since there was a shortage of science teachers and I was going to teach science, and that didn't work.

"I didn't go to school to get out of the draft; I went because that's what I wanted to do. The military was just not in my game plan. I hadn't even thought about it until I got my friendly greetings. You just go do what you have to do.

"The day I was supposed to walk across the stage to get my diploma in May, I was low-crawling as a private E-1 at Fort Bliss, Texas."

After boot camp and AIT (advanced individual training) at Fort Dix, N.J., McClure accepted an invitation to attend Officer Candidate School, and received a commission as second lieutenant in the infantry in April 1969.

"I didn't know anything about the military. I knew where Fort Sam Houston was in San Antonio. I knew where Lackland and Kelly and Randolph (AFB) were, and that's about it. I had no clue, and no one gave me any guidance.

"When I got to Fort Bliss, they said, 'You're qualified for OCS.' I said, 'Whoop-de-do, sign me up.' Three-hundred and twenty-five dollars a month sounded a lot better than ninety-dollars a month pay as a private. I also decided they were drafting me for one reason, and that was to send me to Vietnam. They didn't care how I came back, but I did. I figured I'd have a little control of my own destiny if I got my commission.

"I went to my first duty assignment up at Ranger camp in April, then jungle school, and was getting shot at in Vietnam in September, my second day in the jungle. I got real baptized; real quick."

McClure says he remembers that first operation – a jungle sweep for which he received a Bronze Star for valor – very clearly.

"It was the first operation I was assigned to as a platoon leader. We got air-mobiled into a clearing, which was about two feet of water, sloshed through that and saw enemy sandal tracks on the trail. We called them 'Ho Chi Minh sandals.'

"We moved forward and had to stop. The company commander and the rest stayed there, and they sent me and my unit about three hundred and something meters away for the night-time location. All of a sudden, at daybreak, all hell broke loose.

"There was gunfire down where the others were, and so we immediately picked up and re-joined them. We assessed what was happening, and they sent me out to sweep the area, and there were some bunkers. We went to go check that out, and my scout said he saw them (enemy) running off, and I had my (M)-79 men throw some rounds in there. Then, when we went on in, there was a campfire, hot rice strung across the ground, so you could see they were on the run, and a frog laying there, cut open, with the heart still beating.

"I called my platoon sergeant, and I said, 'Bring all our gear and come up and join us.' All of a sudden, gunfire erupted, and we returned fire. Within a few minutes, we had everyone in the world calling me, and I didn't even know my own call sign.

"They came in with mini-guns and dropping napalm. Dirt clods and sticks were hitting around my head. But we swept the area and did OK. I just had one guy wounded. He got shot in the arm, but it was a clean wound. He retired as a New York firefighter and then as a dentist.

"It's kind of frightening, but you don't have time to think. You have to take care of business. I was on two radios, talking to everybody in the world, having all this going on, and trying to get a handle on the situation.

"After that, everything else was minor. There were a lot of skirmishes, but they were all minor, compared to that one."

McClure came home from Vietnam in 1970, spent some time at Fort Rucker, Ala., and then joined the Army Reserve. He earned a master's degree in guidance and counseling, and wound up settling in central Texas – Harker Heights, specifically – when he heard about a job opening at Central Texas College in Killeen.

He held a number of different positions at CTC and retired as an associate dean for guidance and counseling emeritus after 37 years. He also served for several years as adjunct faculty at the University of Central Texas (now Texas A&M University-Central Texas), where he taught a graduate course for counseling/psychology majors.

So, even though his dream of a career in education was interrupted early by Uncle Sam, he kept the faith and everything worked out in the end.

"When I came back to Fort Rucker, they said, 'What the hell you doin' here? Are you an aviator?' I said, no, and they said, 'Well, we don't have a job for you, so you're going to have to go out and find one.' This was my first time on a military post, and I had no clue about what went on. But being a P.E. major, I just looked at them and said, 'Do you have a gym I could take care of?'

"I ended up being assistant special services officer. I had a good time. I ended up being selected as one of the three captains to be an assistant I.G. (inspector general), and I learned a lot about the Army then, but I'd already decided to get out and go to grad school (master's degree in counseling and guidance).

"Back then, it was a one-year turnaround and then I'd be going straight back to Vietnam as an infantry company commander. I had been shot at enough already.

"There's a lot of unique stories I could write a book on, I think. But I was there to do my job. I looked my soldiers in the eye and said, 'My job here is to go home talking, walking, and breathing, and bring everyone with me.'

"I remember when I went to my first reunion years ago, (and) the guys come up to me and said, 'Sir, uh, we remember your philosophy very well.' They said, 'How come you stayed in the jungle longer than the other lieutenants?' They also talked about how we never had a soldier killed and only about six wounded even though we were in more skirmishes than the other two platoons combined.

"Like I said, I didn't know anything about the Army. I was just like them – I got drafted, but I got commissioned and they stayed enlisted.

"I was just doing what they told me to do. That's how things worked back in those days."

Chris McPhee

Growing up in a family of three boys being raised by a single mom, Harker Heights resident Chris McPhee decided during high school that serving a few years in the military would be his ticket to a college education and a dream career as an architect.

Not long after enlisting, things began to change and those well-intentioned plans soon went out the window.

"We lived in the inner city of Miami – Liberty City – and there wasn't a whole lot of money for me to go to college, so I figured I was going to make my own way," McPhee said. "The military had the GI bill, and so I thought it would be a good way to get some money for college, get a trade, travel the world. I figured I could do four years, and after that, I could go to school and move on with the rest of my life.

"What I ended up doing four years into the military is not what I envisioned doing."

It was 30 days after graduating high school in June 1994 that McPhee found himself in basic training at Fort Leonard Wood, Mo. He went on to spend a couple of years as an administrative clerk, and then one day he attended a Special Forces recruiting briefing.

"I thought, 'Oh, this is interesting.' I get to travel the world – South America, Africa, Asia, Europe. I was like, you know, it would be cool to go work in Africa. I get to teach people; learn skills. I met a couple of people who weren't too successful at the actual assessment portion, but I thought I could do it.

"I was an athlete in high school – all-star wrestler; champion since eighth grade – so when I joined the military, I was in pretty good shape, fitness-wise.

"At that time, I was married already with two kids and one on the way. That initial four-year plan of mine did not include a wife and kids. It was a very transformational time – from being a high school kid to being a husband and father. I started developing my personality; figuring out who I am; learning responsibility. Now, I needed to shape a career. I didn't have an education, and so the military was my place to be.

"I got to Fort Bragg, went to the assessment course, made it, got selected; went to the qualification course, made it. In '96, I was jumping out of airplanes at Fort Benning, Ga., on my 21st birthday."

After the Sept. 11, 2001 terrorist attacks rocked the U.S., McPhee and his Special Forces (Green Beret) team was on its way to Afghanistan, beginning a series of back-and-forth deployments as the war on terror began.

He was excelling at his new career when a traffic stop while he was back home at Fort Bragg turned his world upside down.

"I got a DUI in 1994," McPhee said. "At that time, I was a sergeant first class, so I was facing probably getting kicked out of the Army. When you're in an elite unit, anything they see as a flaw kind of brings you as a person into question. I got a letter of reprimand, but fortunately, I had a good reputation with my peers, my command, and my mentors, and I didn't get booted. A lot of people helped me."

He managed to keep his job and his stripes, but also continued his hard-partying ways. Multiple overseas deployments, being passed over for promotion four times, a divorce, watching friends die, all contributed to a serious depression that kept McPhee going back to the bottle.

"I was drinking a lot. During a deployment, I lost a buddy to murder-suicide. Another roommate of mine, I watched him eat his last meal, and at three in the morning, they called me and said, 'Hey, you're the only one here from this unit, (so) you've got to go I.D. his body.

"I had just seen this guy eating crackers and chicken salad, excited about where he's at, and now I'm looking at him in a body bag the next morning. There was a lot of drinking going on during those next 48 hours.

"If anybody knows the military, everything is centered around drinking. When we're happy, we drink. When we're sad, we drink. When soldiers got issues, they drink.

"On Jan. 1, 2013, I woke up, walked my dog like I normally do in the morning, and it was just like, 'You know, I'm just sick and tired.'

"At that time, I had been going to work, leave at lunch time, don't come back. Get off Friday and get a bottle. And then I almost got another DUI. That scared the crap out of me, because I knew my mentors and all the folks that were rooting for me had said, 'If you do this again, we would have to put you under the bus. There is only one lifeline, and if you mess it up, then it's on you.'

"I'm walking my dog and I said, 'You know what? I need to change my life.' Some little voice came to me and said, 'You need to stop drinking.' So I went in the house and dumped out all the booze that was on our island, and that was day one. I've been alcohol-free now for seven years, nine months, 13 days.

"A lot of great things have happened over the years since coming to the realization that I had an issue, and if I didn't fix it, I was going to destroy myself. If I hadn't done that, I probably wouldn't be here right now."

After earning a promotion to master sergeant, McPhee retired in 2015 with 21 years' service. He worked for a while as a military intelligence analyst in Virginia and was once offered a position with the Secret Service, before starting his own company, Green Beret Media, which provides photography and video services. He also started The Team Room Talk Podcast, which features special guests and a wide variety of topics related to society and culture.

"If I had to narrow it down, it's real talk with real people," McPhee said, adding that his podcast can be found on iTunes, Pandora, and iHeartRadio. "That's it. My goal with each episode is to entertain, educate, or provoke emotion."

Life today is good, says McPhee, a 45-year-old father of four and grandfather of one who moved to Harker Heights recently with his wife, Kristen, a soldier currently on active-duty and stationed at Fort Hood.

People who guided and mentored him through both good times and bad are a huge part of his success, McPhee said, and he has the following advice for anyone else who might be struggling:

"I would say to any soldiers out there – I would say to anyone on this planet – if you don't have a mentor, you're lost. If you don't think you need one, you're wrong. Period.

"If you even have a book that you read and you follow what that person has written, you're good. Or if you watch YouTube videos and that motivates you to do good things, then you're good. But if you're just one of those people who says, 'I'm not going to listen to anybody; I'm going to do this all by myself,' you're lost.

"You're doomed for destruction.

"Mentors. I stack 'em up. I have, like, 10 of them. I tell people, 'If you want to battle me and you don't have any mentors, it's not going to be a fair fight.'"

Chris Scrivens

Retired tanker Chris Scrivens remembers vividly the day his and two other M-1 Abrams tanks came under heavy fire in Iraq, setting the engine compartment ablaze, and causing what turned out to be a devastating injury that went undiagnosed for years.

"It was April 4, 2004," said Scrivens, a 43-year-old retired staff sergeant who lives on Fort Hood. "We went into the city of Najaf after we got word that the current al-Sadr had moved his forces down to that region.

"We were jumping city to city, chasing terrorists, and this battle was going on its tenth day. We were unable to approach Najaf from our side, because there was a bridge that had been daisy-chained with IEDs. Another unit from Charlie Company was able to approach from the opposite side of the city.

"When we discovered how many IEDs were actually on that bridge – it was enough to take out a tank and land it on the riverbed below – we had a scout go in, crawl along the ground (and) take out about three of them. The next day, we went in. Took one of the EOD's robots and had him go across the bridge, cutting each one of the IED cords. The last one, the IED blew up, taking out the robot – poor guy didn't make it. I found a few wheels later on, but that was it.

"We went into the city, fought all day and into the late evening. Sporadic fighting. At that time, our hands were tied. We could not launch our main gun rounds. We worked with the Blackhawks and Apaches to zero in on a target, which was a group of about ten terrorists holding up in one of the college buildings there in Najaf.

"Late that evening, we got approval to launch main gun rounds. There were a number of individuals around a table enjoying tea, and a few minutes later, after we launched main gun rounds, there was no longer a tea party.

"We proceeded back outside the city, and the next day we went back in. That's when we were told to see how far we could go. One of those 'thunder runs.' There were three tanks. At that time, in a company of 14 (tanks), we were down to four operational tanks. We had taken in that much battle damage, as well as mechanical issues that had to be fixed.

"Within 30 minutes, each tank had taken about thirteen RPGs, AK-47 rounds … then we got the call to turn around and exit the city. The first tank got hit; the second tank had a name on the gun tube – we name our tanks after we qualify 'em – of 'Bada Bing.' The RPG-29 went into that gun tube, knocked the 'Bada' out of one side and the 'Bing' out of the other, and the round stayed inside the gun.

"Before getting back over to that nice, friendly bridge that had all those IEDs, my tank got hit and the engine compartment caught fire. My tank commander told me the tank's on fire, and I told him I wasn't getting out until it was out. We were still getting shot at, and when we got over to our side of the friendly bridge, our lieutenant at the time put out the fire.

"We all took a deep breath and said, 'Good job. We didn't die.'"

Unbeknownst to him or anyone else at the time, the concussion inside the crew compartment from that fiery blast damaged Scrivens' jaw, causing progressive deterioration over time that led to a series of major dental issues and reconstructive surgeries.

"That's what the docs traced it back to," he said. "It was a concussion wound that could not be seen at the time. Basically, the lower jaw had receded an inch over a year.

"I had to have my face reconstructed. I wore braces, had 16 teeth removed, eight implants, and a partial in the front of the mouth. In 2010-11, when I should have been going to Afghanistan, I had to stay back for two massive cosmetic dental surgeries."

Originally from Rockmart, Ga., near Macon, Scrivens joined the military's delayed entry program while he was still in high school. He served two-and-a-half years in the National Guard, then went active-duty. During his 22-and-a-half years in the service, he spent 13 years in Germany; multiple tours at Fort Knox, Ky., as an instructor; a stint as a drill sergeant at Fort Jackson, S.C.; and finished his career in 2017 at Fort Riley, Kan.

The father of three says he is proud of his military service and would do it all over again.

"I come from a very small town – graduating class of 141. I did not see myself remaining there. There was not enough life for me there; not enough for me to grow as a person. So, I looked at joining the Army as a challenge, an adventure, a growing opportunity, and a way to get out of a small town.

"When I first went in, I thought I'd spend a few years, then get out and move on to something different. But as time went by, I enjoyed what I was doing. Different things make you decide to stay. You decide to stay for family; you decide to stay because you like the current location you may be in. You like where you are and what you're doing, so you decide to stay a few more years.

"I think that's how it normally goes. People go in thinking they're going to do their contract – do four to six years – but over time, they learn what kind of impact and what kind of lessons there is to learn being a soldier, and they enjoy their job. They enjoy how much influence they have on other people around them," said Scrivens, who works now for a local remodeling contractor, and also serves as quartermaster for VFW post 3393 in Kempner, and for the Combat Veterans Motorcycle Association 23-5.

"The Army has changed over time, but I can never complain about my time in the military. The biggest thing for me was .... even after you leave the military, your name lives on. It lingers around motor pools; it lingers around older sergeants, and people who met you while they were young.

"Your impact in the military is widespread. You don't just do your job. If you're a good NCO, a good sergeant, your impact generally is felt across a huge spectrum of people. Even people who may not know you; they know of you. I believe that young soldiers have to learn that they do not know everything. Individuals above them with time and knowledge are an example to them – right and wrong. If they're smart, they'll take the good aspects of all and decide what kind of soldier they're going to be.

"I think the military is still a good option. A lot of young people need purpose. They need to grow up within a structured environment, where other people have learned life lessons. Your mother and father can only teach you so much, that you're willing to accept. At a certain point, you always tend to tell them (your parents), 'I know better than you; I know more than you.'

"In the military, that doesn't apply. The military has its own cycle, and those people who have been in a while know that cycle. So, you telling them 'I already know this and that' does not work.

"I learned that you can mop a motor pool in the rain – you can sweep a motor pool in the rain," Scrivens said, laughing at the memory of him being assigned such character-building duties. "And you can absolutely wash a tank in the rain."

Bill Shine

Vietnam veteran Bill Shine says he learned a number of valuable life lessons during his time in the military, and some of those came when he was assigned as an aide to a legendary general who first came to fame in World War II.

"While I was at Fort Jackson (South Carolina), I got selected to be an aide to the commanding general, who was Lt. Gen. James Hollingsworth," Shine, a longtime Harker Heights resident, said. "He was a graduate of Texas A&M, a famous World War II guy, and the third-highest decorated Army officer in U.S. history.

"He was tough as nails, and I learned a lot. After I went to work for him, he said, 'Bill, I'm not going to tell you what time to come to work, but don't let me ever beat you here.' So, at six o'clock every morning, I was sitting in that office.

"Sometimes it was pretty close, but I don't think he ever beat me."

Shine is a bit of a rarity for central Texas, having been born in Temple and then living nearly his entire 75 years in the Killeen-Harker Heights area.

"Yeah, we're a rare breed," he said. "I've spent all my life in Killeen and Harker Heights, except for the time I was in college, a year I worked in Dallas after college, and then when I went into the Army.

"So, about an eight-year gap there, but other than that, I've been right here."

Shine graduated from Killeen High School in 1963, then headed 300 miles north to Lubbock to attend Texas Tech University, where he earned a bachelor's degree four years later in business administration, went to work for a year, and then received his draft notice from Uncle Sam.

"Since I already had my college degree, I could volunteer for OCS (officer candidate school) before I had to report for the draft," Shine said.

"So, I got accepted into OCS, and still had to go to basic training, which I did at Fort Dix, New Jersey. That was in January, which was cold as heck, especially for a Texas boy who hardly even had a jacket.

"I went to Fort Benning, Georgia, for OCS. I was there for eighteen weeks, and from there went to Fort Jackson, South Carolina, to be a basic training officer, training draftees. Most of those guys were going straight to Vietnam."

In late December 1969, Shine found himself on a plane headed to Vietnam, where he would soon be leading fellow troops into combat during a one-year tour.

"It was pretty rough," he said. "When I was on the plane going over there, I wondered how I was going to react (leading troops in war). The first firefight I was in … your adrenaline kicks in, (and) you rely on your training, and you really don't think about it. You just do what you have to do.

"There was a lot of action, but you find out a lot about yourself. I think I was a good leader. I got back alive, and most of my men – not all, but most – got back alive."

One of the many memories he has from his time in Vietnam was when a U.S. helicopter pilot spotted what turned out to be a makeshift enemy motor pool in the jungle. First Lt. Shine and his platoon headed that direction, and within a couple of hours located an array of 33 cargo trucks, small pickup trucks, and several Land Rovers – one of which was brand new with only 730 kilometers (453.6 miles) showing on the odometer – along with a stash of vehicle parts including bearings, brake shoes, axles, transmissions, batteries, pistons, a large generator, barrels of gasoline, and cases of motor oil.

Throughout the area, there were underground sleeping quarters with electricity, a mess hall with live chickens and pigs, an estimated 50 tons of bagged rice, first-aid facility, and recreation area with a ping pong table.

"That was during the Cambodia invasion, before the 1st Cav went in there," Shine said. "I was a platoon leader, and we're the ones who actually found this NVA motor pool.

"Remember Gen. (Robert) Shoemaker (former Fort Hood commander)? He was assistant division commander of 1st Cav at that time, and he flew in and looked at all the stuff, and he said he wanted all the Land Rovers to go back to our firebase. So, it took us about two days – we had to blow down trees and everything – and we made a caravan and drove those Land Rovers back.

"They used them on the firebase to carry stuff around. I always thought that was pretty neat that we had Land Rovers to drive around."

After his tour ended in December 1970, Shine came home and left the Army after three years' service. He might have stayed in longer, but an opportunity to go to work for his dad was too good to pass up.

"I was a first lieutenant and in less than a month, I would have been promoted to captain. But my dad had a Goodyear tire business here in Killeen (Shine Brothers Goodyear), and I had an opportunity to come back here and go into business with him.

"That was in 1971, so I did that until 1998. I had a good opportunity to sell my business and I thought, 'You know what? This opportunity may not come around again,' so I decided to go ahead and take advantage of it."

After nearly three decades selling tires, and his wife of 52 years, Jean, running the highly successful Shine Team Realtors, Bill decided it was time for a change of pace. He went home one day and announced he was getting out of the automotive business.

"I came home and told her, 'Hey, I made an agreement to sell the Goodyear stores.'

"She said, 'What do you plan to do now?'

"I said, 'I'm going to work for you.'

"She said, 'What are you going to do?'

"I said, 'I'm going to empty the trash, run the vacuum cleaner, wash the cars …'

"She said, 'No, it doesn't work that way. You're going to get your real estate license, and you're going to go to work.'

"So, that's what I've done ever since then. Actually now, I'm more in the operations part of the office. I go to work every day and try to do enough to make it so they'll keep giving me a (pay)check."

The father of two, grandfather of two, and longtime civic leader, was among eight area military veterans honored recently with a Congressional Veteran Commendation for outstanding community service.

Over the years, Shine has served on the Central Texas College board of trustees, chairman of the Heart of Texas Defense Alliance, board of directors for First National Bank Texas and Fort Hood National Bank, civilian aide to the Secretary of the Army. Other honors he has received include the Killeen Independent School

District distinguished alum award; Fort Hood Good Neighbor Award (with wife, Jean); Outstanding Civilian Service Medal from III Corps and Fort Hood; and the Roy J. Smith Award (with Jean) from the Greater Killeen Chamber of Commerce.

His Army career proved to be short-lived, but Shine says he is proud of his service, and credits the experience he gained there with contributing to his later success.

"I learned discipline, of course, and I definitely learned how to take correction. I really enjoyed my time in the Army, but because of the opportunities I had back here, I decided to get out. I certainly wouldn't have minded staying in.

"It was definitely a big part of my life," said Shine, who received a Bronze Star with oak leaf cluster for bravery in Vietnam, along with an Air Medal with 'V' device. "I learned how to lead people, and how to treat them. The biggest thing is, you always put the soldiers first. If you were going through a chow line or something, you make sure the soldiers get fed first."

Tracey Brown Greene

Tracey Brown Greene grew up in tiny Farmville, N.C., graduating from high school in 1996 and going on to study computer programming at a local community college.

She was serving in the Army National Guard but had no intention of making the military a career.

Then, things changed.

"I just decided going into the Army would be easier," said the 43-year-old Harker Heights resident, who has lived in the central Texas area since 2008.

Growing up with two older brothers, a dad who served in the U.S. Marine Corps, and a mom who was in the Army Reserve, the then-20-year-old had some idea what to expect when she got off the bus for basic training at Fort Jackson, S.C., but even so, she was not fully prepared for what awaited her.

"It was a lot different back then," Greene said. "Now, they don't do them the same way as they did back then. I haven't heard a soldier in the last ten years say that they were screamed at, or they took anything from them.

"We got off the bus and there was lots of screaming and 'Hurry up; hurry up!' The bags that we had, they took those and we had to go through them and take out very little.

"They said, 'You can leave your civilian clothes in the bag – you don't need those. We're going to give you everything you need.' Then, they locked our bags up, and issued us everything else.

"I thought I was kind of prepared for it, because I had been around soldiers all my life, but it was like, wow … it was kind of funny, really but we knew not to laugh."

After boot camp, Greene went to AIT (advanced individual training) at Fort Lee, Va., and then to her first duty station as a member of the XVIII Airborne Corps at Fort Bragg, N.C., where she worked in logistics as a motor pool clerk.

"I did a lot of reports. One of my bigger jobs at that point was turning in equipment. All the equipment that had to be retrofitted had to come through me to get fixed. I did a lot of turn-in and receiving."

From Fort Bragg, she was assigned to duty in Germany, then back to the States to Walter Reed Military Medical Center in Washington, D.C., followed by an assignment as a supply sergeant at Fort Meade, Md., issuing equipment for the only combat camera company in the Army.

She is a veteran of the war in Iraq, with two Middle East deployments: one in 2009-10 and another in 2016-17.

The mother of two boys and a girl retired from Fort Hood as a master sergeant with 22 years' service in 2018, and now runs a glass and screen installation company for residential and commercial properties.

She enjoyed her career in the service, although it was difficult at times being a female in a mostly all-male organization. The U.S. has an active-duty Army today of 476,000, including around 74,000 women.

"There were problems sometimes," Greene said. "Females seemed to have more of a problem just with you being a female, and so they wanted to be hard on you. Males were more, 'I don't feel that you can actually do this job,' and so I prefer to hit on you.

"Looking back, the highlight of my military career is the soldiers. The actual impact that I had on my soldiers. I had soldiers who are now officers and warrant officers, senior NCOs. They have gone off to do great things, and the ones who got out of the Army are brokers, own their own companies, and are doing amazing things.

"When I was able to talk to soldiers, or the ones that were having issues, and we had meaningful conversations, and I was able to help them, those are the highlights.

"The Army is still a good choice. You just have to be aware of your surroundings at all times and use it for what it is. If you plan on being a career soldier, use it to be a career soldier. If you plan to be in for a minimum amount of time for an enlistment, use that time to the best of your ability. Use the Army for what it is.

"It's a tool – a stepping stone for your next destiny."

Earl Williams

Nolanville resident Earl Williams served 30 years in the U.S. Army, achieving the highest non-commissioned officer rank of command sergeant major, and went on to a successful career selling life insurance after he retired from the military.

Then, tragedy struck his family back home in Wilmington, N.C., and changed the direction of his life.

"My sister came down with Alzheimer's," the 74-year-old father of two and grandfather of three said. "I brought her out here to live with me, and I took care of her until she passed away.

"I just said there's more to life than being out there making money. Family is a big part of my life, because I've been traveling and away from home for so long. Her husband had passed away and she didn't have any children, so my wife and I said we were going to make sure she is taken care of.

"When you're taking care of someone with Alzheimer's, it's a 24/7 job. You can't set your own schedule; you have to go by their schedule. After going through that experience, that's when I decided I wanted to do more to help other people, especially our veterans."

Williams, who grew up in a family of four boys and three girls led by a father who worked building roads for the state of North Carolina, started volunteer work after his sister's death and has become so involved with various volunteer organizations that he was one of eight veterans honored recently for outstanding community service by the Congressional Veteran Commendation Program.

He joined the military right out of high school in 1965, following in the footsteps of two brothers, one who also served in the Army and another who served in the Air Force. At that time, joining the Army all but guaranteed at least one combat tour in the controversial war that killed more than 58,000 U.S. troops.

Despite the obvious danger, Williams said he never had second thoughts.

"None whatsoever. I didn't even flinch. I knew that was part of it when I joined. I knew what I was getting into and what could happen, so it did not bother me at all."

After basic training at Fort Jackson, S.C., and advanced individual training at Fort Knox, Ky., Williams was assigned to his first duty station in Germany. In April 1969, he was a sergeant serving with the 101st Airborne Division in Vietnam.

"I flew in helicopters as an area observer," Williams said. "Our job was to go out in the jungle, find the enemy, report to the troops, call in air strikes. It was a pretty dangerous job, because you got shot at every day."

On June 24, 1969, Williams was aboard a helicopter flying over triple-canopy jungle terrain, scouting for enemy troops, when the aircraft was rocked by enemy fire.

"When we got shot down, everything went blank for a while.

"I can remember some of it, but not all of it. It's hard to explain. Everything was happening so fast, and then all of a sudden, everything seemed to slow down. I guess my training kicked in, and after we were on the ground, it was like I was doing things in slow motion.

"Medivac was called in. They had to drop what was called a jungle penetrator to pull the wounded out, and then the medics carried us to the field hospital.

"The infantry we were supporting was coming up, and they kept them (enemy) from capturing us. Had we not crashed that close to the unit we were supporting, we were possible candidates for prisoner of war.

"I know it had to be a couple of hours or better, because they had to call in the Medivac and we had to make sure the area was secure for the helicopter to come in and over, and not get shot down."

Williams, who received a Purple Heart after suffering severe injuries from shrapnel, spent a week or two in a field hospital. He nearly earned a ticket back home, but eventually wound up back on combat duty.

"I can't remember how long I was in the hospital, but it was long enough for them to evaluate me on whether they were going to ship me back to the United States, or another place they were sending soldiers to recuperate.

"I wanted to stay, because I had a brother in Vietnam at the same time I was there. He was in a different part of the country with the Americal Division, and I wanted to get back into action. I wanted to give some payback."

Was he able to deliver that payback?

"Oh, yeah," Williams said. "Most definitely."

He finished his tour in April 1970 and came home to Fort Hood. Over the next 25 years, he served with such famed units as the 1st Cavalry Division, 1st Armored Division, and 2nd Armored Division, earning promotions all along the way until he reached the top of the heap at E-9 (command sergeant major), a somewhat rare achievement he attributes to "hard work and being dedicated to doing my job, and doing it properly."

In November 1989, Williams was squadron command sergeant major for the 11th Armored Cavalry Regiment assigned to the Out Post Alfa checkpoint along the border between then-East Germany and West Germany, when the heavily guarded border was opened and the historic Berlin Wall began to be dismantled. In July 1990, he became regimental command sergeant major for the 11th ACR in Fulda, Germany, where he remained until coming back to the States in 1994.

A year later, Williams decided to hang up his boots, retiring from the 1st Cav's 2nd Brigade. He went to work for MetLife for a while, then stepped down to care for his ailing sister, who died in 2004.

Since then, Williams has become more and more involved in a number of veteran service and community organizations including the Area Veterans Council, Disabled American Veterans, VFW, American Legion, and Military Order of the Purple Heart, Hood Howdy, and Rotary Club.

"Caring for my sister was life-changing," he said. "If anybody ever has to take care of a parent or relative, it really changes – for most of us; maybe not all – where you place your values. It put me more into wanting to volunteer and help others.

"I had saved my money, and my wife worked in civil service as an education counselor at Fort Hood, so I decided there was no use in me being greedy – I can give back to the community. That's when I decided that I wanted to do more for the veterans. I started volunteering for the Armed Forces E9 Association, and I'm still volunteering there.

"I just enjoy so much giving back."

While his myriad volunteer efforts occupy a considerable amount of his time, Williams, who has been married to wife Doris for nearly

54 years, also enjoys visiting his kids and grandkids, and even heading out to the lake now and then for a little relaxation.

"I make time now and then to get down there and do some fishing – mostly, I wind up just feeding the fish," he said, laughing. "They love to see me coming, 'cause they know they're going to get fed."

Guadalupe Lopez

Guadalupe Lopez was haunted for decades by some of his experiences serving with the infantry in Vietnam (1969-70), but two memories are especially painful.

"It was probably about May," the retired Army sergeant major said, during a recent phone call from his home in Killeen. "I was flying with the colonel in his command and control helicopter. We got the word that an engineer team clearing the roads of explosives had hit a mine. We were close, and the colonel said, 'Let's not wait for a medivac; let's go on in there.'

"It was two engineers that got hit, and when we got there, one of them was on one side of the road, and the other was next to the chopper. One of them was dead, and the other was missing part of the hands, and missing part of the legs. We tried to pick him up, but it was hard to grab a hold of his body.

"He was almost naked from the explosion, and his body was just limp – like a balloon filled with water. So we got a poncho, rolled him onto that, and put him on the chopper. Then we took him to the military hospital, and a couple of minutes later, after we were back up in the air, it came back on the radio that he was dead on arrival.

"That one stuck with me for a long time, but the second incident really affected me."

Lopez was NCO in charge at his tactical operations center and he noticed a couple of officers preparing to leave the base emptyhanded – without carrying a weapon. He ran after the pair, but they shrugged him off.

"I had a captain who was getting ready to come back to the States," Lopez said. "He was getting an officer to replace him, and he wanted to take him out to show him around. A small observation helicopter came in, and they were walking out of the TOC. Both of them didn't have their rifles.

"I ran out and hollered at them, but they said, 'No worries, Sarge. We're OK.'

"I went back in and was monitoring the radios, and we got a call that the LOH (light observation helicopter) was hit. The pilot said, 'I'm hit. I'm going down.'

"We got a call from an Air Force forward observer plane that was close by that said he had located the helicopter, with one person underneath it. He also mentioned that he saw another person moving away from the chopper, and another one moving in the opposite direction.

"What happened was, the chopper went down, (and) the new officer died on impact. He was thrown out, and the chopper landed on top of him. The pilot had always carried his weapon – a sidearm and an M-14 – and the recon team located him and brought him back.

"They couldn't find the other officer. Two days later, we got a report (about) a freshly dug grave with hands and feet sticking out. They went back out and recovered the body.

"That," Lopez said, pausing as his voice choked with emotion, "has stayed with me all this time. We used to work together, side by side, and it hurt me a lot losing him. For quite a while, I blamed myself for not running out there with the weapons. It took me almost thirty years to get it out of my head."

Now 82 years old, Lopez has lived in Killeen since he retired from the U.S. Army in 1980. A native Texan, he was born and raised down south in Laredo, and joined the military when he was 16 years old.

His father had been out of work for a year, and his cousin told him he wanted to join the service.

"We were like brothers, and in September, we went to the recruiter. They kicked me out of the recruiting station in Laredo. He said, 'No, you have to wait until you're 17.' My cousin was accepted, and he was going to El Paso for basic training. We went to San Antonio, and we told the recruiter there that I was 17. He said, 'No problem. Sign here.'

"Four years later, they caught up with me, and they tried to kick me out for falsifying the records. I was getting top secret clearance at that time, and they started digging into my official records. But I was doing very well in the military, and the battalion commander wrote a good letter excusing me."

After basic training at Fort Bliss in El Paso, Lopez was assigned to Fort Jackson, S.C., for advanced training, then shipped off to Alaska. He came back to Fort Lewis, Wash., got out of the Army for a while, attended college and got married, then re-enlisted a year later, serving two tours in Korea, one tour in Vietnam, and a stint as an advisor at the U.S. Embassy in Lima, Peru.

In October 1962, he nearly saw his first combat as part of the U.S. invasion force during the Cuban Missile Crisis, a short-lived but historic confrontation between the former Soviet Union and United States over Soviet missile deployments on Cuban soil.

"I was with the first unit that left Fort Hood," Lopez said. "It was an armored cav reconnaissance troop. We were on a maneuver at Fort Hood, and they cut that off halfway through. They brought us back and told us to prepare to reload again tomorrow, which was on a Sunday. We thought it was just another drill, you know, loading up on the railroad (cars).

"Once we were loaded up, they told us, 'You got two hours to take care of whatever you want to. We're headed somewhere.' They didn't want to tell us where.

"I had to get a hold of my wife and tell her we were leaving, and we didn't know where we were going.

"That night, we left by train and went to Dallas that night. From there, we went to Fort Stewart, Ga. They had a set of civilians going from one end of the tracks to the other, fixing all the vehicles and loading them up with ammo. That's when we knew we were going somewhere for action.

"We loaded our tracks (personnel carriers, tanks) onto these LSTs and once we got out (to sea), they told us that night to turn around and go back.

"We got back and landed at Hollywood-by-the-Sea, north of Miami, and we stayed there for two to three weeks, at the Gulfstream race track. From there, they brought us back to Fort Hood."

After he retired, Lopez went to work for Killeen's Central Texas College as director of military training on Fort Hood. He stayed there 20 years and then retired. Now, the father of four – who is remarried after losing his wife of 51 years, Hortencia, to cancer in 2009 – is writing a book about his time in Vietnam, 'The Battalion Staff Duty Journal: 2nd Battalion, 14th Infantry in Vietnam – One Soldier's Story.'

Tremendous pride is evident in his voice when he talks about his military career. Once a soldier, always a soldier, Lopez says.

"I love the Army. It is in my heart, and I will never forget it. You can't tell me anything bad about the Army. Every time I see a soldier, I say thank you very much. You're a one-percenter.

"What I say to young soldiers is, they need to understand that the work in the Army is the same as on the outside – it's a profession. If they come in, they need to apply themselves to whatever they're supposed to be doing.

"If they don't like it, get out after three years, or four years, and try to find something they can fit into on the outside. But if you stay in, do your best. I made sergeant major when I was 19 years in the Army. I made all my promotions faster than my peers, because I was always studying, volunteering, going through a lot of schools.

"I always tried to do my best."

Jon Burrows

Retired Air Force Col. Jon Burrows has an impressive resume that includes bachelor's and master's degrees from the University of Oklahoma; a law degree from the University of Texas; 20 years' service as a central Texas county judge; and 30 years' service in the military.

In spite of those and a long list of other accomplishments, however – including hanging out with legendary UT football coach Darrell Royal and equally legendary musician Willie Nelson – it was an unforgettable couple of days back in 1969 that produced what the 74-year-old Temple resident calls his "claim to fame."

"I played saxophone in the OU band," Burrows said. "I was first chair, and I got a call from my saxophone instructor, and he said he had gotten a call from a group that was coming in to play in the college concert series.

"They needed a saxophone player, and he couldn't do it, so he asked would I be interested in doing it. He said they might pay me something, and I thought, 'Sure, sure.'

"The band was The Beach Boys. When I saw the music, I thought, 'Well, this is a piece of cake.' It was just background music – whole notes and half notes. I played with them there at OU, and they said, 'We're gonna go up to Omaha. Can you come up there with us and play up there?' I said, 'Sure.'

"I think I got paid union scale, which was thirty-five to forty bucks – a lot of money at that time. So that's my claim to fame … playing two concerts with The Beach Boys. They were very nice guys. A good bunch."

Burrows was born and raised in tiny Tipton, Okla., the son of a farmer and power transmission line repairman. The youngest in a family of one boy and two girls, he graduated in 1964 as high school valedictorian and promptly enrolled in college, where he studied instrumental music education, played in the OU band, and served in the Air Force ROTC.

After completing his master's degree, Burrows went on active duty to fulfill his military obligation and maybe see the world.

He did not get far.

"I remember the dream sheets you fill out (for) where you'd like to be assigned," Burrows said. "I had not been anywhere except for southwest Oklahoma and OU. So, my first choice was Germany; my second choice was Japan; my third choice was the East Coast; my fourth choice was the West Coast.

"In their infinite wisdom, they sent me to Wichita Falls, an hour-and-a-half from home. My travel pay was seven dollars and fifty cents."

Burrows served four years on active duty, including stints as an education and training officer at Sheppard Air Force Base, and then Offutt AFB in Omaha, Neb. He left active-duty and despite having other career goals throughout college, decided to go to law school.

"I'm not sure what the real reason was. I thought I was going to be a high school band director. One of my roommates was a business major, and he was looking for a job. I was going into the Air Force, so I wasn't concerned about that, but I saw there was a world of difference between the jobs he was being offered compared to the jobs school teachers were being offered, so I thought maybe I needed to re-think my plan.

"I started looking at what my alternatives were. I was always a good test-taker, so I took the law school admissions test, and did real well on that. Also, the graduate business school test. I was looking for maybe a combination MBA and JD program, and actually got admitted to the business program at the University of Chicago and Northwestern University, but those didn't match up with the law program.

"UT didn't have the MBA program at that time, but it was a top law school and I thought, 'Well, I can go as an in-state resident and this is a name school,' and it turned out to be a good decision."

That was 1973. Richard Nixon was president; the Watergate scandal was unfolding; the war in Vietnam was winding down; Secretariat won the Kentucky Derby; and Burrows was spending time in Darrell Royal's living room in Austin listening to Willie Nelson play guitar.

"My wife's dad and Darrell were the two youngest brothers in the Royal family. They were very close. We stayed with Darrell and (his wife) Edith for a couple of weeks at the beginning of law school.

"Darrell used to have what he called picking sessions at their house, and we got to go to some of those – sitting in the living room with about twenty other people while Willie Nelson played the guitar. I didn't know who Willie Nelson was, but I found out quickly."

It was also then that Burrows joined the Air Force Reserve, where he spent the next 26 years, serving after law school with the JAG (Judge Advocate General's) Corps at Bergstrom AFB, and later as senior JAG officer for the Reserve at Randolph AFB in San Antonio.

In 1976, Burrows moved to Temple, where he practiced law for 20 years, and became an active part of the community, serving in a variety of positions including: board of trustees at Temple College; chairman of Temple Chamber of Commerce; United Way of Central Texas campaign chairman; president of Temple Founder Lions Club; and chairman of the Temple Civil Service Commission.

"When the position of (Bell) county judge came open, I decided to throw my hat into the ring," said Burrows, who retired as county judge in 2018 after two decades in office. "I had no idea what I was doing, but it kind of matched what my involvement with the community had been. I knew enough about it to know that you had an opportunity to be involved in all aspects of the county, and to work with people from all aspects of life. I thought it would be a very good opportunity for me to serve the community, and it was. It was better than I anticipated.

"I had a contested primary and a contested general election that first time around, and my last two terms, I was unopposed. If you're ever running for office, and you have the opportunity to run opposed or unopposed, always pick unopposed. It's a lot easier and cheaper.

"County judge is elected countywide, and you work with four commissioners who are elected by precinct. To get anything done, I just needed two people to agree with me, and most of the time, all five of us agreed. I was very fortunate I had outstanding commissioners my entire tenure. Some of the best and smartest people I knew."

He counts construction of the new Bell County Justice Center as one of his proudest accomplishments as county judge, along with expansion of the county jail and improvements to the Bell County Expo Center.

The father of two and grandfather of four says at least some of the credit for all his success can be traced back to his time in the Air Force. He and Katie enjoy traveling when they can, and every fall, when Oklahoma and Texas square off in Dallas for the famed Red River Shootout football game, he cheers for the burnt orange.

"I married into the Texas athletic program, so even though I was in the band at OU, my allegiances are to the University of Texas. I pay my alumni dues to both schools, but I root for Texas."

John Footman

Retired U.S. Army Staff Sgt. John Footman was drafted for military service in 1967, a couple of years after high school, and soon found himself patrolling the jungles of Vietnam.

Enemy small-arms fire caught him in the lower left leg one day, but he finished his 1968 tour, came back to the States, and then was assigned to duty in Germany. Nine months later, he volunteered to return to the war in Southeast Asia.

"I wanted to go back because the mission wasn't finished," said Footman, a 74-year-old Harker Heights resident. "I never told anyone that I volunteered to go back, but one of my brothers cornered me one day and said, 'I want you to tell me the truth, and don't lie to me.'

"I said, 'What's going on?'

"He said, 'You volunteered to go back?'

"I said, 'Yeah, but don't say nothing to nobody.'

"He told me – and excuse my language – he said, 'Well, brother, you better bring your ass home.'"

Footman was born and raised in Tallahassee, Fla. He knew the jungles of Vietnam was his destiny when he received his induction notice June 1, 1967, but he was ready, willing, and able to go do his duty.

"I was glad I got drafted," he said. "My cousin had volunteered, and he tried to get me to volunteer, but I didn't think they'd take me. I got my draft notice two weeks after he had volunteered.

"My mother knew it was coming. She knew I wanted to go in the Army. She just said, 'Son, take care of yourself.'"

Basic training was at Fort Gordon, Ga., followed by advanced individual training at Fort Polk, La., then it was on to combat in

Vietnam. During his 20-year career, he also served at Fort Benning, Ga.; Fort Eustis, Va.; Fort Polk; Fort Hood; and a total of 12 years in Germany.

After retiring from the military in July 1987, he and his late wife, Kathleen, who died last October, returned to central Texas. Kathleen managed a mobile home community, while Footman worked for a time at the Veterans Administration, then took a job as a jailer with the Bell County Sheriff's Department. After that, he worked security at Scott and White Memorial Hospital, where he stayed for nine-and-a-half years before having major surgery and deciding to retire permanently.

A past commander for the Harker Heights chapter of the Military Order of the Purple Heart, and the Military Order of the Purple Heart Department of Texas, Footman now serves as chief of staff for the organization's Region V. He also helps out with the local Wreaths for Vets group and is an officer with the Harker Heights Public Safety Commission.

He recently was honored along with eight other area veterans for ongoing community service as part of a Congressional Veteran Commendation program started six years ago by Rep. John Carter, R-Round Rock.

"I'm involved with all the veterans' organizations around here. With my wife gone, it helps keep my mind occupied, and I like helping soldiers. When soldiers call me and say, 'Do you know anybody that can help me?' I find out what their problem is, and I send them to DAV 29 (Disabled American Veterans Chapter 29 in Harker Heights), and those guys really know how to help out.

"I've helped soldiers that were wounded in Iraq and Afghanistan. I've helped them get their Purple Heart; their disability benefits.

"I like helping soldiers because didn't anyone help me when I retired. I found out everything on my own. These young soldiers getting out now – they need help. And it takes us older veterans to give them the support they need. They tend to give up too easily, and I tell them not to give up fighting for their disability."

Now considered 100-percent disabled, Footman received two Purple Hearts, two Bronze Stars for valor (one with Oak Leaf Cluster), three Army Commendation Medals, and a Combat Infantryman Badge.

One Purple Heart was for that leg wound during his first tour in Vietnam; the second recalls that explosion that nearly killed him after he volunteered for a return trip.

"When I got back there, they sent me to the 4th Infantry Division, and I was talking to the commander. He said, 'What are you doing back here, Sergeant?'

"I said, 'Well, you've got a mission, right?'

"The first sergeant said, 'That's the man we need in platoon.' Come to find out I was the only one in the company that was a Vietnam returnee."

Footman eventually wound up being assigned to the 5th Infantry Division (Mechanized), and 30 days after that, a massive explosion rocked his armored personnel carrier.

"I hit a 60-pound mine. The concussion caused me to not be able to feel anything from the waist down. I couldn't feel nothing. Three doctors said I'd never walk again.

"I had been laying in bed for about 15 days, trying to figure out how I'm going to go home and explain it to my mother. I told my buddy one day, 'I've got to move my feet.' So, I moved my toes, and I said, 'Wow, man, I think I can get up.' He said, 'John, don't get out that bed.'

"I got up and when I stood up, I kept falling. I mean I hit that floor hard. I'll never forget this little major coming in there and she said, 'What are you doing?' I said, 'I've gotta walk.' The doctors were shocked. They gave me a walker, and I walked with a walker for about a week, then I told my buddy one day, 'You know what? I'm through with this walker.'

"I put it to the side, and I've been walking ever since.

"When I went back to the unit, the first sergeant said, 'I don't believe it.' I had a month-and-a-half left (in his tour), and he said, 'You're not going back out to the field.' I said, 'Top, I want to go back out with my soldiers.'

"He said, 'Sgt. Footman, what did I say?' I said, 'OK, but I've got to do something.' He said, 'I heard you worked in maintenance before.' I said, 'Yeah, I've done maintenance.' So, they sent me to the motor pool for my last month-and-a-half, but I kept going on R and R. I went to Sydney, Australia, three times; went to Hong Kong once."

He still has a little trouble with hip and back issues, and some of the mental scars from those long-ago combat experiences – and the treatment he and many other soldiers received back home – continue to haunt him even today, Footman says. Nevertheless, he is proud of his military service and has no regrets.

"People spit on us, called us all kind of names. It wears on you when somebody calls you a baby killer. You have to have a strong mind and a strong constitution.

"The thing that sticks with me right now is … a little (Vietnamese) girl was walking toward us. She was probably six or seven years old, and she had something strapped to her. My squad leader told me if I didn't shoot her, he was going to shoot me.

"When I shot her and she hit the ground, she blew up. Things like that happened a lot.

"When I look at little kids today, it still bothers me. If anybody tells you they went to Vietnam, and it didn't bother them, they're lying to you. I still have flashbacks sometimes. I wake up in a cold sweat, and it was over 50 years ago.

"But I enjoyed my time in the military. The good Lord saved me over there. If I had to do it again, I'd do it all over. I really would.

"And now, I like helping these young soldiers coming into the Army. A lot of them never had a father; some of them never had a mom. Those guys need our leadership."

Adam Butler

When his older brother got shot through the jaw and neck during combat in Vietnam, Adam Butler knew exactly what he was going to do after graduating high school in Columbus, Miss.

"I was kind of upset," said Butler, now 71 years old and a resident of Copperas Cove for the past 30 years.

"He got shot during the Tet offensive (one of the largest, fiercest campaigns of the Vietnam War) in January 1968. It went into his jaw and came out through his neck on the other side. He survived, but it tore up everything in his mouth.

"After that, I decided to join, more or less to go over there and find out who had shot him. Everybody said, 'You know you're going to Vietnam,' but that didn't bother me. I wanted to get over there as quick as I could."

Butler was still in school when his brother was wounded, but he did not waste any time enlisting. He graduated and just a few months later (September 1968) found himself flying UH-1 Huey helicopters through the jungles of Southeast Asia.

"I went to boot camp at Fort Polk, Louisiana, then Fort Wolters, Texas," Butler said. "I went through school there with the little Mosquito helicopters, and then I went to Fort Rucker, Alabama, and on to Vietnam.

"We flew mostly chief of staff missions, mostly for the commanding general of 1st Cavalry Division. Sometimes I had to fly troops around, if they were short on pilots, somebody was out sick, things like that, but mostly it was chief of staff missions.

"In a way, it was pretty safe duty. We flew over hostile territory, carrying documents and people back and forth from Saigon. I was never shot down, but we did almost get blown up one time.

"We were out at an LZ (landing zone) area, and they had a lot of incoming (enemy fire). We went out there to check on an engineer unit, and they started shooting rockets in on us. I remember that really well. It was something else.

"At that time, I was flying Gen. (Robert) Shoemaker (a former Fort Hood commander). The helicopter was on the ground; the blades were still turning; and the rockets were coming in. We had to get him out of there.

"Vietnam was not a fun time … well, we did have some fun times when we got a break. When we went down to the coast on R and R, we enjoyed ourselves. Just rested, went into bars, drank beer, went out to the beach, and had a good time.

"Yes, we talked to a few girls, too. Those French-Vietnamese were some beautiful women."

Butler went on to serve a total of 20 years in the U.S. Army. He retired at age 38 as a staff sergeant.

"When I came back from Vietnam in March 1970, I was a CW2 (chief warrant officer). They had too many officers at that time, so if you wanted to re-enlist and come back in, you had to come back in as enlisted.

"I was supposed to go to III Corps (at Fort Hood), but they sent me over to the 2nd Armored Division. They said, 'Well, we've got to figure out what we're going to do with you guys.'

"The next day, they called me down and said they had a job for me. They wanted to send me up to Fort Sill (Oklahoma) to be over an M-60 machine gun firing range. They said, 'They got sergeants up there who will show you the ropes. All you're going to be doing is showing the ROTC guys how to fire the weapons. You just have to sign off on everything.'

"So, I went on up there. When I got back, I got in my car and went down to a 7-Eleven in Killeen, and that's when Capt. Elliott showed up. He drove up the same time as I did. When I got out of the car, he said, 'Butler, where have you been? We've been looking for you. You're supposed to be at III Corps.'

"He got me out of the 2nd Armored, and so then I was at III Corps Headquarters for a while."

Butler left the military in 1988 and attended some classes at Central Texas College in Killeen before a 95-mph windstorm in May 1989 caused hundreds of millions of dollars in damage to aircraft at Hood Army Airfield and Robert Gray Army Airfield, and he went to work doing helicopter maintenance and test flying.

Now fully retired, the father of three and grandfather of two has been married to wife, Cenora, for 35 years. He spends time these days helping out around the house and working on cars. Sometimes, he pitches in to help friends with mechanical work, and sometimes he turns a few wrenches on his prized 1976 Chevy Monte Carlo, which he bought brand new at the old Connell Chevrolet in Killeen.

On occasion, Butler also has been known to get together with a friend or two and rent a Cessna airplane for a little joy-flying.

"Airplanes are a lot easier to fly," he said. "Helicopters are not easy – except when you get into the Cobras and the Apaches. But the rest of them will wear you out.

"A lot of people say, 'You ought to be proud of yourself. So many places you've been and things you've done.'

"I say, 'Well, I don't really talk about it too much.' My wife, we were married for about a year before she even knew I could fly. Her dad always said, 'No, no, he didn't fly in the Army.'

"One time, she went out to West Fort Hood to pick me up, and she saw me get in a helicopter and take off, and she said, 'I seen him.'"

Butler and his 73-year-old brother live 700 miles apart now – one in Copperas Cove and the other in their hometown Columbus, near Tupelo, birthplace of Elvis Presley – but they keep in touch. He made it to the area of Vietnam where his brother was wounded, but never learned much beyond that about what happened.

"I went up to where he told me he was, but I didn't get a chance to get out there. In '69, it was still too hot (dangerous) to go out in areas like that.

"He went to Japan, and they shipped him to the medical center in San Antonio. They repaired his mouth, and guess what he became when he got out of there? An E-6 drill sergeant at Fort Bragg. Ain't that something?

"He got out shortly after that, because they didn't want him to bring his family up.

"He doesn't like to talk much about the military. He was kind of upset because he had wanted to stay over in Vietnam. I said, 'Well, it's good that you came back and you're still alive.'

"I enjoyed my military career. It was a lot of fun."

Randy Dyer

Boot camp was a breeze for 56-year-old Copperas Cove resident Randy Dyer when he spent six weeks at Fort Jackson, S.C., until it was time to ship out for the next phase of his military career.

"There was a battalion of privates graduating from basic training, who are then going to travel to different spots for their advanced training," Dyer said. "You probably remember that during that summer (1981) – it was the Reagan presidency – the air controllers had a strike and because of that, we couldn't leave.

"The biggest airport near Fort Jackson (S.C.) was essentially shut down. There was 150, 200 guys that were at the transitional disembarkment point, or whatever they called it, and we all had tickets to go someplace, but nothing to get there on.

"We were all standing around outside this building with nowhere to go. About midday, these big trucks and a forklift came in. They dropped off a bunch of aluminum bleachers, so we had something to sit on. We sat there the whole day. It ended up the next morning there was a convoy of private buses taking us to Atlanta, which is a major hub. That was quite an experience."

Dyer, who works now as a security analyst for the information services security engineering department at Baylor Scott and White, grew up in Shawnee, Kan. He was minding his own business, going to school, working part-time, and serving in the Army National Guard, when his life took an unexpected turn.

"My dad was a religious zealot, and we didn't get along, so I had moved out of the house," Dyer said. "I had joined the National Guard, and I went to basic training the summer between my junior and senior year. Came back home and decided that living at home was muy no bueno (no good).

"I was in the Army now and you couldn't tell me anything anymore. I had gone and done all kinds of basic training stuff – shot machine guns; threw grenades. You want to get all over me about whatever?

"See ya. Bye. I grabbed my stuff and left."

With just months remaining before he graduated, school officials suddenly notified Dyer there was a problem with his record. Supposedly, he owed a considerable amount of money for books he had failed to return several years before, and they were telling him to either pay up or get out.

"They kind of forced me to drop out," Dyer said. "They said I had not turned in my books from my freshman year, and I had until that afternoon to produce the books or pay for their replacement.

"I was working the night shift at Sambo's (restaurant). In the evenings, I would bus tables, and overnight, I would cook. I had not planned on leaving high school, so I was trying to figure out what to do, and one of my teachers – who I had the utmost respect for – chewed me out and gave me a ration of crap because I didn't come and talk to him about it."

That friendly teacher helped Dyer get approval to take the GED exam, followed by reporting for advanced individual training (AIT) at Fort Leonard Wood, Mo., then back to Kansas to attend classes at the local community college.

He was doing well in school and excelling at his National Guard duties, but after a time, Dyer decided full-time, active-duty Army might be a good idea.

"I wanted to do more," Dyer said. "In the National Guard, they have a different hierarchy. In a lot of instances, if you want to get promoted, you've got to wait for someone to leave or die.

"I was stuck in a position where I was an E-5 sergeant, and I was sitting in an E-4 specialist slot. I was over-ranked for the slot, and they couldn't over-rank me another one, so I had to find another job. In our battalion, they had a rations section, so I was in charge of two squads. I would get the orders from the battalion mess sergeants for the foods they were wanting to get, and I had to pre-stage my orders for all the food, receive it, issue it, and do turn-ins, paperwork justifications … the whole bit.

"Then they said, 'We can't keep you in the slot, unless you're M.O.S.-qualified. You're doing a bang-up job; you should be an E-7.

You can go to this National Guard school training one weekend, once a month, for a year-and-a-half. When you get done, then you'll be M.O.S.-qualified, and we can promote you to E-7.'

"I said, 'I don't want to go to school for a year-and-a-half. That's ridiculous.'

"They said, 'Well, you could go to the active-duty Army M.O.S. school. You'll be going up there as a sergeant, and you'll be around all those privates, and they got women up there. You can't fraternize because you'll get in trouble.'

"I was married by that time, and so I wasn't interested in anything like that. I had been making active-duty E-6 money, which was a lot better than the liquor truck I was driving. So mama and I talked about it, and we decided I was going to go active duty.

"Put in my notice. Left my job. Went down and had an appointment with a recruiter. He said he was going to take care of me. He took me down and we're at the very last station, and they guy says, 'Where's your grade determination paperwork?'

"I said, 'The what?'

"He said, 'We can't process you without it. Because you're in a Reserve component, coming in to an active-duty component, they must evaluate your rank before you come in. You can't come in as an E-6, if they don't approve it.'"

A few weeks later, Dyer was on active-duty, but with a rank of E-4. The corresponding decrease in pay might have been an issue, but in this case, it worked out OK.

"I sucked it up and went on active-duty as an E-4 with maximum time in grade," Dyer said. "They sent me to school at Fort Lee, Virginia," Dyer said. "Gorgeous country. Used to run P.T. in the forest – it was awesome."

Overall, Dyer spent 18 years and change in the Army. He would have stayed longer but for a serious knee injury that eventually qualified him for a medical discharge.

"I was riding my bicycle in Korea and there was a Korean taxi driver that ran me off the road," Dyer said. "I hyperextended my right knee and tore my ACL (anterior cruciate ligament). So the rest of my time in Korea, I was all messed up.

"They kept telling me that I had a really bad sprain, and here's some Motrin. Don't run, jump, or march for x number of days.

"The first guy at the orthopedic clinic back at Fort Hood was like, 'Oh, you ain't got no ACL.' I had an operation, and it did not go well. So they did another one. They saw some other problems in there, and they said they could do a third surgery, or they could medically discharge me.

"I had already been on two years of profiles, surgery recoveries, physical therapy, and I didn't want to have to do that anymore. So I got out."

These days, the father of two works remotely from home due to the COVID-19 pandemic. He enjoys practicing the guitar, learning new computer applications, and crocheting large blankets to give as gifts to friends and family. He and his wife of 30 years, Laura, also raise adopted greyhound racing dogs.

He has no regrets or resentments about his military career and the way it ended. He also looks forward to the world getting back to "normal" after the upheaval that began early last year.

"I know there are folks out there who like to bash on the military, but I don't harbor negative sentiment about the Army," Dyer said. "I'm not bragging when I talk about sleeping in a mud hole. I walked on top of the Imjin River when it was frozen a foot-and-a-half thick, on the DMZ in February in Korea.

"The Army helped me provide for my family, and that's a huge thing. I was able to have access to health care. We had a solid roof over our head. If I had to get up at two in the morning and be in the office by three, OK. That's what you do. We all have things we have to do, and if we don't, a lot of times they'll get rid of you and find somebody else who will.

"I'm hoping in the near-term that we will be able to give each other a handshake, pat on the back, and maybe a neighborly hug. Personal contact and social interaction is part of what has brought us above the other animals.

"Chimps can take a stick and it's called a tool, but I've never seen a chimp build a suspension bridge.

"I think it's really important that we have the ability to interact. We're social creatures, and to be denied that is really horrific. And all of the small businesses that have collapsed under the weight of this (pandemic), or are struggling, I can't even imagine.

"This is going to be a huge economic impact for the country for years. If they keep stepping on the throat of the economic engine of our country – small business – what other option is there? We need to get beyond that and get back to being social creatures, sharing the love and lifting up our fellow man."

P.C. Santiago

Retired Master Sgt. P.C. Santiago remembers searing heat on his back and blinding light in his covered eyes as he witnessed atomic bomb explosions in the early 1960s at the U.S. government's Pacific Proving Grounds in the Marshall Islands atoll.

After World War II, the United States conducted nuclear testing on the Marshall Islands for more than a decade, and Santiago served there for one year as a young draftee doing building and other maintenance.

"On the day of testing, they marched the entire company to the shore," said Santiago, a longtime Killeen resident who served much of his 30-year military career at Fort Hood. "When we got there, they gave us dark glasses – the kind that you can actually look safely at the sun – and they said, 'Everybody turn around; do an about face.'

"We did that, and they started counting: 'Ten, nine, eight, seven, six …'

"When that thing went off – we must have been 15, 20 miles away – you could feel the heat. It was as if somebody had a torch at your back. It was unbelievable. It was so bright that even with special glasses, you could not look at the explosion.

"We had to have a Geiger counter around our necks to measure how much radiation was in the air. After that, they march you back."

Now 85 years old, Santiago was born and raised in Puerto Rico. He dropped out of high school in the 10th grade and moved to New York City, where he was working in 1955 when a draft notice arrived in the mail. He was 19 when he reported to Fort Dix, N.J., for basic training.

"I didn't mind boot camp at all," Santiago said. "To me, it was a chance for another adventure; a new experience. A lot of the kids that were there with me were long-haired city kids. They were not used to, for example, the rain, the mud … and, for some reason, I loved it.

"I didn't mind marching ten miles in the mud or the snow."

His first duty station was Fort Hood, followed by a stint at Fort Knox, Ky., then on to the Marshall Islands, and over to Germany in 1962. His M.O.S. (military occupational specialty) was changed to medical specialist, and he was assigned to work at a burn center, a job he called "the most rewarding work I've ever had."

In 1966, Santiago was sent to Vietnam, where he served one year as a medic. He saw plenty of things that still haunt him from time to time, but one incident in particular still brings tears to his eyes.

"We were going one time by the riverbed," Santiago said. "We saw an American bulldozer moving sand, and there were two Vietnamese kids using it as a ride. When the dozer would move, they would jump off the tracks.

"This kid was, like, eight years old, and he couldn't jump fast enough and he fell under the tracks. The driver stopped, but his body practically exploded. He was making gurgling sounds. They would not let us get close, so we stood there and watched that. We watched his last breath.

"You get used to seeing men badly hurt. You always see a few dead people, but that was the hardest thing I ever saw. Seeing twenty dead people would not have affected me like that did – watching a little boy just being crushed. You go through many different feelings. You think about what you could have done. The worst part is that it won't leave your mind."

Another lasting memory from Vietnam involves giving medical attention to local prostitutes who provided their services to the American soldiers.

"I was protecting my troops," Santiago said. "I would fill up my (medical) bag with so-many shots of penicillin, and we would go to the place where the girls were. I would tell the interpreter to ask if the girls had noticed anything wrong with their body: drainage, burning, color change. If they say, yes, I would go ahead and give them a penicillin shot.

"That didn't happen a lot, because we tried to teach some improved hygiene and things like that. In other words, you don't just go and boom-boom somebody and then two minutes later, somebody else."

After Vietnam, where Santiago was awarded a Bronze Star for "exceptional duty performance," he came back to Texas and went to work at the military burn center in San Antonio and was promoted to E-7 (sergeant first class). He later worked at a burn center in Japan, and then served at Fort Hood from 1972-85.

He became chief medical NCO for the 1st Cavalry Division and served under 10 commanding generals by the time he retired with 30 years' service. A licensed vocational nurse with an associate degree in general studies from Central Texas College, Santiago went to work for 15 years for the Veterans Administration before he decided to retire for good in 2001.

The father of three, grandfather of five, and great-grandfather of three more, says he is proud of not only his military career, but also the years he spent after that helping other vets at the VA.

One of the ways Santiago keeps himself busy and involved is writing letters to the editor of the Killeen Daily Herald. He has a binder full of clippings from what he figures must be hundreds of his letters that have appeared in the paper.

"There are things that bother me," he said. "When you go to a place like Vietnam, you realize what being poor really is. You have people over there eking out a living from nothing. Living in the United States – are you kidding me? This is like living in heaven.

"America is about honesty. It's about being who you really are. If you are a cowboy and feel comfortable wearing cowboy boots, what are you going to do with a pair of oxford shoes?"

Bill Evans

Bill Evans was preparing to fight to the death after his reconnaissance airplane was shot down in Vietnam, but as he and two shaken-up buddies hunkered down and waited for enemy troops to appear out of the surrounding jungle, they saw one of the most beautiful sights any combat soldier can see.

"The pilot, co-pilot and I all agreed we were not going to be taken prisoner," said Evans, a retired U.S. Army master sergeant who served two tours in Vietnam, 1964-65 and 1968-69. "We weren't going to go to the Hanoi Hilton (notorious prisoner-of-war camp), so we were prepared to shoot it out with them."

Evans, a Dallas native who has lived in Copperas Cove for 37 years, served as a radio operator for most of his 28-year military career. He was on a mission that day to detect and locate enemy radio transmitters when his airplane came under fire and crash-landed.

After that, there was nothing to do but wait for rescue and hope for the best.

"The pilot was great – I will always remember Mr. Wilkerson. He come down and went into a dive, then flopped the plane into the jungle," Evans said.

"I had all the classified stuff in the aircraft, so I destroyed all the equipment I had, and we're sitting there with rifles, waiting. We could hear the noise (around us), and we're getting ready to go into a gun battle, and then all of a sudden, here they came out of the woods.

"We were ready to blast them, but then they started saying, 'No fire! This is friendlies. No fire.' It was what the Army called a LRP (long range patrol). They saw the plane go down, so they came to us and provided a perimeter around the plane until they could get another plane to pick us up.

"When they started speaking English, we said, 'Come on in.'

"We wanted to kiss 'em."

Evans, an only child who was born at Parkland Hospital in Dallas, moved to California with his parents during World War II. He went to high school in Lompoc, near Vandenburg Air Force Base, but dropped out of school when he was 16 and joined the Army.

"My family moved to California to work in the war-machine factories, and they stayed after the war," the 87-year-old said, relaxing in an easy chair recently at his 10-acre home. "I was having problems at home with my stepfather and I wanted to get away, so I joined the Army. I forged my birth certificate so I could get in and reported for boot camp in 1950 at Fort Ord (California).

"It was a piece of cake for me. I enjoyed it. I enlisted to go to the Far East, but they decided they needed radio operators, so they sent me to Fort Monmouth (N.J.), where I learned Morse code and stuff like that, and become a field radio operator."

After that, Evans was on his way to Korea and his first taste of combat.

"The Korean War started the day I got out of basic training," Evans said. "I got out July 25, 1950 and it was December 23rd, I believe it was, when I landed at Pusan (South Korean port city). I was a rifleman, but I was also a radio operator.

"The thing I remember most from Korea was trying not to get killed. I was there four months. That's when they found out that I was a 16-year-old kid, and they didn't want the publicity of a 16-year-old getting killed – and I certainly didn't want to get killed.

"I got a call from the company commander wanting to see me. I fell out and went to headquarters and he said, 'Sit down. You're going home, son.'

"I said, 'I don't want to go home.'

"He said, 'Well, you lied when you came into the Army, didn't you?'

"I said, 'Oh, yes, sir, you got me.'

"At that time, I was in the process of getting a top-secret clearance, and they found out through that. I wanted to stay with my unit, but my mother had signed for me and so she had the final say. When she signed, the Korean War had not yet started, and now she didn't want me in the war. She changed her mind, so I was out.

"They put me on a plane and flew me back to Monterrey, California. I was thinking, 'What am I going to do now?' Soon as I landed, I went with a buddy of mine to Manville, New Jersey, and worked for a company called Johns Manville. Then we come on down the coast and worked for Douglas Aircraft. I just did whatever struck my mind at the time.

"After a while, I got bored because I was doing the same thing all the time. That's when I said, 'I'm just going to go back in.' I thought my mother would try to talk me out of it. But she didn't. She said, 'It's your life. You do whatever you want to.'

"That made me mad, so the next day I went down and re-enlisted."

Evans, who served with the Army Security Agency for much of his career, retired from the Army while he was stationed at Fort Hood. He had no idea what he was going to do next.

A friend of mine was the chief probation officer in Gatesville and he asked me what I was going to do. I said, 'I don't know, s--t. I guess I'll get me a tin cup and my overcoat, and sell pencils at the main gate.'

"He said, 'How would you like to be a probation officer?"

"I said, 'I never thought about it.' I knew I never wanted to be a cop, and that sounded awful close to being a cop – you wear a badge and carry a gun. But I took that job and did it for fifteen years. Worked in Lampasas. Then I retired from that and went fishing and hunting the rest of my life."

Twice divorced, Evans lives now with his three dogs: a German shepherd named Arko, and two yellow Labrador retrievers trained for hunting, Shirley and Ruby. Although he has an associate degree in general studies from Central Texas College and a master's degree in criminal justice, Evans never earned a high school diploma, and that always bothered him.

A few years back, he decided to do something about that.

"I got to thinking about it, and I have a friend who still lives in Lompoc, and I said, 'Will you check with the high school and see if they'll give me my high school diploma, now that I've got two college degrees?'"

School officials agreed and five years ago, he flew out to California, put on a cap and gown, and participated in his high school commencement ceremonies with 300 young students.

"I just felt like I was missing something," Evans said. "I'm happy now."

These days, Evans moves a little slower than he would like, and so he is not able to hunt as much as he once did. He still shoots the occasional dove, but mostly loans Ruby and Shirley out to friends and family members who hunt. Other than that, he watches TV, noodles around on the computer, and feeds a small herd of deer that come calling at the back door every morning and evening.

"I don't cook, and I don't want to kill something I won't eat," said the father of three and grandfather of four with another on the way. "I finally just got to the point where I woke up one day and I said, 'You know, I'm not mad at these animals anymore.' I still like to hunt birds, but I've gotten so old that I can't swing (a shotgun around) fast enough to shoot 'em.

"Life is good. I'm 100-percent disabled, so the VA gives me a check every month for that, tax-free. I get my military retirement; retirement from the state from when I was a probation officer; and I'm drawing social security.

"I'm making a fortune and I don't have anyone to spend it on. I think you should be born rich and then work it down from there. When you're young, you have to watch your money, and when you get old like me, everything is paid for.

"I saw the world. I did lots and lots of things that I'm glad I did. I went to places like India and Pakistan, Germany and Japan, Korea and Vietnam. I'm just tickled to death."

Conclusion

Some pretty amazing stories, don't you think?

As always, when I started this new series of veteran profiles for the Killeen Daily Herald newspaper, I had no idea how it would go. Some vets do not particularly enjoy talking about their time in uniform, and a handful of ex-soldiers that I contacted turned down my request for an interview.

Most all of them agreed, however, and the feedback I have gotten has been tremendous and gratifying.

I love telling stories about people, and the soldiers I know are some of the finest people around. No matter what you think of our "war machine" and the politics that go along with it, remember that the cogs in that machine are just regular folks like you and me who make tremendous sacrifices and do it willingly.

Sure, there are lots of life-long benefits that come with a successful military career, and that is no doubt a motivating factor for enlisting. But, if you ask me, someone who spends a significant amount of time in a military uniform earns their keep.

Is there a more important and high-stakes job?

So, the next time you see someone in one of those camouflage uniforms and a pair of combat boots, remember that they are not some sort of robot soldier created in a laboratory or something. There is a regular ol' person inside those BDUs who hurts and worries and cries and bleeds just like anyone else.

The difference is they are a regular flesh-and-blood human being who is called upon to do incredibly brave and dangerous things.

There was a time when I did not believe soldiers who told me they were looking forward to deploying overseas and going to war. I figured they were just repeating the "company line," so to speak. Giving the responses they had been trained to give.

Who in hell would really be excited about going someplace where people are trying to kill you?

At some point, I realized they were telling the truth.

They really do look forward to it, although most downplay the enormity of their task, saying they are ready to go and "do their job."

Think about it this way – if you are an athlete and all you do is practice, practice, practice, and never get to play in a real game, you probably are not going to be a happy camper.

Same thing with soldiers.

They want to go play in the game.

To the men and women in these stories, I say thank you. Thank you for the interviews and thank you for your service.

Hoo-ah!

A Small Favor to Ask

Thank you for reading this important book. I hope you found it helpful, useful or meaningful in some way.

If you did, please take just a moment to write a brief review of it on Amazon. Your reviews not only mean a great deal to me, but they also help other people find this book. You never know, your review might touch someone's life. How cool is that?

Go ahead, take a minute now and write that review. It's important.

Go to the book listings page on Amazon. Scroll down to the reviews section.

Thanks in advance.

About the author

John Henry Clark lives in a one-stoplight central Texas town, roughly halfway between Austin and Waco. A graduate of the University of Houston, Clark is an award-winning journalist, author, athlete, freelance writer, personal trainer, photographer, musician, artist, husband and father who has more than a dozen published non-fiction books, including his bestseller, *Camino: Laughter and Tears Along Spain's 500-mile Camino de Santiago*, which chronicles two of his three backpacking treks along the historic pilgrimage across northern Spain. A tireless seeker, researcher and questioner, John has written a number of other fascinating books dealing with the human experience, from tragedies to triumphs and more, including his first published title, *Finding God: An Exploration of Spirituality in America's Heartland;* the riveting, *Depression Blues,* in which he talks about how he learned to overcome a lifelong struggle with depression and anxiety; and *Finally Fit: It's Never Too Late to Achieve a Dream,* an inspiring success story of beating the odds and learning to live again.

For more on John and his books, along with information on his writing services, visit: http://www.johnhenryiii.com

To find out more about John Clark Fitness and his senior fitness program and personal training services, go to: http://www.johnclarkfitness.com

johnclarkbooks

More from John H. Clark III

(NOTE: The following is an excerpt from John's follow-up tribute to military veterans, "Courage in Camouflage II," another collection of true stories told by the men and women in uniform who lived them.)

James Wilkins

James Wilkins grew up dirt-poor during the Great Depression of the 1920s and '30s, quitting school after sixth grade to go to work and help his parents support his seven brothers and sisters.

When the Japanese launched a surprise attack on the U.S. naval base at Pearl Harbor in Hawaii on Dec. 7, 1941, the young man from West Texas who was picking cotton and beans as he approached his 21st birthday knew immediately what he had to do.

He enlisted in the Army and was soon on his way to basic training at Fort Sill, Okla.

"My country was in the war, and they needed me," Wilkins said recently from his home in Nolanville, where he has lived since 1980. "Back then, you had to be 21 to enlist, so I had to get a telegram sent from Mom to allow me to go in.

"It was January 1942. All I remember is that the war had started, and I went and joined. That's it."

After boot camp, Wilkins was on his way to Fort Hood (then known as Camp Hood) for more training, out to California, then over to New York, where a troop ship ferried him to England to prepare for crossing the English Channel and eventually heading into combat in Europe.

By this time, he was a gunner on an M18 Hellcat tank destroyer, a fast-moving and highly lethal machine deployed to find and eliminate the German army's famed Panzer tanks. A fellow crew member, Arlee McCracken from California, was his tank commander. The two had first become fast friends while they trained at Hood, then brothers-in-law after they met a pair of attractive sisters during a weekend trip to Waco.

"These two were in training at Fort Hood, and they went to Waco for ... weekend activities," said Wilkins' son, Steve, who lives now with his father to keep an eye on things and help out around the house. "They went into this area and there were these two redhead sisters.

Arlee McCracken said, 'Hey, I like the little short one over there with the red hair.'

"Dad said, "Well, I like the other one.'

"Uncle Arlee ended up marrying my mom's youngest sister almost immediately. Dad tried to marry the other sister – my mom – almost immediately, but she said, no. She said, 'I'll be here making airplanes in Fort Worth when you get back. If you're still interested, that'll be fine.'"

The elder Wilkins interjected:

"She said, 'We'll get married, and I'll get pregnant, and you'll go over there and get killed, and I'll have another baby to raise.'"

Steve continued:

"So Arlee became my future uncle – I wasn't born yet – and then Dad got out in 1945, came back and married my mom. So they became not only tank commander (McCracken) and gunner (Wilkins), they became brothers-in-law.

"They had never met before they were put together in that tank."

It was D-Day Plus 26 – July 2, 1944 – when Wilkins and McCracken finally crossed the Channel and rolled onto the famed beach at Normandy, France, to join the push toward Germany. At some point, Arlee was hit in the jaw by sniper fire and sent home, leaving Wilkins as tank commander. Just as he and his crew neared Berlin, the Germans surrendered and the war was over.

Wilkins was credited with five Panzer "kills" and awarded a Bronze Star and Silver Star. He also participated in the liberation of prisoners from a number of German concentration camps, including the notorious Auschwitz complex in occupied Poland.

At one time, he and his fellow crew members lived for 18 months in their M18, taking a grand total of one shower before they reached Berlin.

"It didn't matter – we all smelled the same," Wilkins said, smiling.

"I went to several prison camps. All we did was shoot around them, open the gates, and the people came out. Let me tell you – they came out. They started walking home."

After the war ended in 1945, Wilkins came back home, that cute girl he met in Waco kept her promise, they got married, started raising a family, and spent the next 58 years together before Esteene died 10 years ago.

With millions of ex-GIs home from the war and looking for work, jobs were scarce. After about a year, Wilkins decided his best bet would be to go back into the military. He re-enlisted and spent another 22 years in the Army, retiring in 1965 as a sergeant first class.

He tried his hand at building houses in the Dallas area for a while, then came back to central Texas and landed a job in civil service at Fort Hood, where he spent the remainder of his working years.

Wilkins, who has three children, eight grandkids, 10 great-grandkids, and two great-great-grandchildren, recently celebrated his 100th birthday. He has slowed down a bit over the years, his eyesight is failing him and his hearing is not what it used to be, but with Steve's help, he proudly hangs on to his independence as much as possible.

He enjoys watching TV, listening to audio books – westerns are his favorite stories – and playing Mexican Train Dominoes on Friday nights.

"I can get around here just fine," he said. "I can walk – in fact, I think I can run …

Steve interjected:

"Well, let's talk about your knees, then …"

Dad conceded.

"I have to watch when I'm walking, because the right one (knee) will give up after so many steps. I have to be careful. I've got a walking thing right there, and another one out on the porch …"

"He won't use them," said Steve.

"Those are for old people," Wilkins added.

One of his sisters lived to be nearly 105, so chances are fairly good that Wilkins will be around long enough to blow out a few more candles on birthday cakes.

He is proud of his military service and says his life has been a good one.

"I enjoyed it, and I'd do it again. Like I said, my country was at war and they needed me. Simple as that.

"It's amazing that I've lived this long. I don't know why that is, but I do know that God has always taken care of me.

"The whole world has gone crazy, but as far as I'm concerned, it's still a good Christian country. I just hope when I'm gone that the world knows I was a Christian."

johnclarkbooks

www.ingramcontent.com/pod-product-compliance
Lightning Source LLC
LaVergne TN
LVHW100316210926
848480LV00003B/1141
9798996924523